ALSO BY RON LIDDIARD:

BAULKING GREEN: Champion Hunter Chaser

GIN & GINGER:
A lifetime spent farming in the Vale of the White Horse

Four Generations of Liddiard through the male line.

*The author **Ron Liddiard** as a baby, being held by his father, Cyril John.*
Seated: L.to R. *Great grandfather George; Grandfather Ernest Liddiard.*

Ron Liddiard

Gin and Ginger:

A lifetime spent farming
in the Vale of the White Horse

Baulking Books
Forty's Farm, Baulking, Faringdon, Oxon, SN7 7QE, U.K.
Tel:01367 820410

First published in Great Britain 1998 by
Baulking Books
Forty's Farm, Baulking, Faringdon, Oxon, SN7 7QE, UK.

Reprinted December 1998
Reprinted February 1999

British Library Cataloguing-in-Publication Data.
A catalogue record for this book is available
from the British Library.

ISBN: 0-9532578-0-0

© Ron Liddiard, 1998

All rights reserved.

Ron Liddiard asserts his moral right to be identified as the Author of this work in accordance with the Copyright, Designs and Patents Act, 1988.

This book may not be reproduced, in whole or in part, in any form (except by reviewers for the public press), without prior permission in writing from the publishers.

Printed by *Manuscript ReSearch Printing*
P.O. Box 33, Bicester, Oxon, OX6 7PP, England.
Telephone: 01869 323447   Fax: 01869 324096

DEDICATION

To the memory of
my dear wife *Ann*
with whom I was extremely fortunate
to spend 58 years of married life.

Sadly she died on 14th December 1997 before the book was printed. I shall always be deeply grateful for the encouragement she gave me and the peace of mind in knowing that she had read the manuscript to the very last word.

CONTENTS

		Page
Introduction by Candida Lycett Green		9
Acknowledgements		11
Preface		13
1	Ancestry and Early Childhood	15
2	World War I: 1914-18	50
3	Move to Baulking	60
4	School Days	67
5	Growing Up	101
6	Chink of Light	163
7	Taking the Plunge	200
8	World War II: 1939-45	203
9	Post War Period	227
10	Wireless and Telly	264
11	The Sixties	272
12	Ol' Baulking	280
13	Fuller's Earth	288
14	Tailings	294

INTRODUCTION

Ron Liddiard's book is a moving and remarkable work. It is a record of a life well spent and an ever unfolding chronicle of a small farm in this, his well loved corner of Berkshire. What comes shining through is not only Ron's love of home and of memory but above all Ron himself — a true countryman with a background of well settled beliefs, passed from father to son for generations.

Anyone who has ever met Ron will know that wherever you may be, whether it is in the village shop or the pub or standing in the rain in Faringdon Market Place, he will *always* tell you a story; and moreover, it will *always* make you laugh. There isn't a better story teller around and he has shown extraordinary tenacity and energy, as an octogenarian, to put down in words what he's been telling friends and neighbours bit by bit over the decades. Perhaps I shouldn't be surprised for Ron comes from stalwart stock! His grandfather, who had a grocery business in Faringdon, thought nothing of breakfasting at 3.30 in the morning, travelling in a cart behind a fast trotting cob the 11 miles to Swindon, taking the train to Bristol, buying his bulk grocery stores, returning to Swindon, trotting home to Faringdon, mounting his hunter and putting in a full day's hunting with the Old Berks.

Laced into the story of Ron's life, are wonderfully evocative images set in this fine stretch of the Vale of the White Horse, where willows still line the banks of Rosey Brook and the archaic line of Downs will never change. Here, trying his hand at pig breeding, being run away with by a couple of cart horses while hay-making, going to Cricket club dances at Shrivenham

Memorial Hall, Ron describes the Vale my parents knew and later the Vale my brother and I knew too. I went to my first dance in Shrivenham Memorial Hall and I spent my Old Berks hunting life as a child trying to avoid jumping Rosey Brook.

The continuance of things, and the layering of memories one over another make me feel stronger when times are tough. Ron has passed on the story of his contented life which can only serve to enrich us all.

Candida Lycett Green
Uffington, March 6th, 1998.

ACKNOWLEDGEMENTS

I wish to record my heartfelt thanks to all those, past and present who have helped in any way at all towards my modest literary effort to describe the story of my life — most of it anyway.

Especially, The Late Hon. Lady Betjeman for writing a letter to me in 1973 *(copied on the following page)* which sowed the seed, so to speak.

To her daughter, Candida Lycett Green, for nurturing the struggling plant when at times it looked more than delicate.

To Patrick Eyre for similar attention prior to the two-leaf stage.

To Suzanna Johnston who told me she read part of the earlier script 30,000 feet above Russia while on her way to visit one of her daughters in Hong Kong. Alas, Suzanna Johnston's help was brought to an untimely end by a serious car accident, from which I am happy to say she is mercifully recovering.

To my daughter, Pam, for hours spent deciphering my handwriting into type and on to disk, and many other helpful things she has done for me during most of the 10 years it has taken to get my efforts into print.

To Wendy Clark for finalising the manuscript into an acceptable double spaced document containing well over 100,000 words.

To Ann Allen-Stevens for her invaluable help with proof reading.

Thanks to all who have provided photographs.

Finally, to Doreen West for introducing me to Val and Graham of Manuscript ReSearch.

Copy of an original letter dated 6 May, 1973, written by The Late Hon. Lady Betjeman to the author, Ron Liddiard.

Harton's Piece
W. Challow
6 May 73

My Dear Ron,
    I stayed last night with Sybil, I was very tired but was KEPT AWAKE till 1a.m. by WHAT DO YOU THINK??
<u>BAULKING GREEN</u>
And I woke at 6.30a.m. and finished it.
    I had seen it here before and glanced at it and knew that one day I would read it through and last night I did.
I really do congratulate you on writing in the sort of way which really grips the reader so that one cannot put the book down. Of course, it makes it extra interesting when one knows so many of the characters in the book, but apart from that it really is extremely well planned, and written so naturally that one is quite carried away. I want a copy for my son in law Rupert Lycett Green which Sybil will buy for me and if you will post it to my new cottage I will be very grateful —
New House, CUSOP, Hay on Wye, Hereford.
    Sybil tells me that Ethel Reade died a fortnight before my recent return from India and that it was a glorious funeral with a real sense of fulfillment. But the O B H will never be quite the same without her flying around the lanes on her old bike. What a gallant girl.
    Next time I am down in these parts I will call and see you and Ann: with love to you both —
    (Signed) *Penelope*
P.S Your next book must be YOUR life with your experiences in the Home Guard etc. etc!
    You MUST write more.

PREFACE

Following suggestions by a few friends who apparently enjoyed reading my modest literary effort about the great Hunter Chaser 'Baulking Green', I wondered if an autobiographic account of the times in which I lived, worked, loved, laughed and cried might be worth recording.

The purpose of writing in the first person is not to illuminate my own participation — this is far removed from deserving any kind of recognition — but purely to use it as a medium for describing an era as I have seen it, in this locality and to talk about some of the people past and present, who have helped to form the minute part of creation's infinite mosaic which has been my life of 84 years to date almost all of which has been spent farming at Church Farm, Baulking.

Memories recalled by my late parents and others long since gone mean that the period recounted would span the years roughly between 1900 and 1997. It could be that these years constitute an era which is on the verge of disappearing and may soon become difficult to recall in the mad rush of present day living. Certainly some of the 'characters' who lived in it will be lost to memory in a few generations and so shall I. The thought may occur — So what? — to this I can only answer 'it happened' and leave the conclusion as to whether it was worth recording to those who read on.

Although not qualified to make particularly knowledgeable observations, I do feel, due to the ever increasing shifting population, essential of course for livelihood and made easy by modern high speed transport; that it will become increasingly difficult to pin a 'character' label on anyone; for the simple reason that most men and women who are looked upon as such attained that description by living and moving around in a fairly confined district for most of their lives and where the population does not change very much either.

Could it be that future generations might come to know their fellows by looking in the right ear to find their registration number and the left ear to ascertain the block and flat number in which the 'character' resides!

It is possible that the last 50 or 60 years have produced more technical, industrial and social change than the world has ever experienced before in such a short time. I think this is certainly true in farming and there is no doubt the two world wars which have taken place within this time span, have been responsible for a great deal of what has happened.

Who knows what further change is in store during the next similar period of time. The world's armouries; stocked with all kind of horrific devices and methods for killing one another — the struggle for space domination — and the increased use of the micro-chip are but three of the many ingredients that will affect the lives of many millions of people throughout the world. Strangely, the die seems to be cast for them and despite their numbers they appear unable to do very much about it — indeed, they may not wish to. "O that men could honour one another and seek the common good" — maybe therein lies the answer to most of the worlds' problems, perhaps the *only* answer — but there, — I had better get on.

So spurred on by a further reading of Lady Betjeman's letter to me, dated May 6th, 1973, reproduced herewith, and the quote from another friend, Patrick Eyre, "Chinese say, the longest journey starts with but a single step" to cheer me on my way, in the words scratched on the door of the old farm privy: "YER TIS".

ONE
Ancestry & Early Childhool

My mother told me it was a hot thundery night when Dr. Hayward ably assisted, I'm sure, by 'Nana' Godwin delivered me into this world in the late spring of 1913 at what was and still is the last house in Faringdon on the Swindon Road. Known then as Coxwell Villa, it stands on a slight rise, with a southerly aspect looking across the Vale of the Uffington White Horse towards the Berkshire Downs some four miles distant, embracing the countryside in which I have had the good fortune to live ever since and come to know so well.

My recollections of Dr Hayward are only faint but I remember my mother saying what a kindly man he was, adding to the thought that some mothers often retain a soft spot for the doctors who attend them during happy events — understandable I suppose! Aren't we men lucky! Confinements in those days mostly took place at home and were the concern of local midwives, district nurses and local doctors.

I have further cause to be grateful to the kind doctor, as he apparently pulled me through a nasty bout of bronchial pneumonia during my first winter. Doubtless not an easy thing to do with no antibiotics and the like to help him. Some years later I remember seeing an old steam kettle in the attic, it had a long narrow spout about two feet long. Wonder if it had anything to do with his success?

Nana Godwin lived to a good age and there must be some people in and around Faringdon even now who in the early days of their lives were cradled in her comfortable trusty arms, lulled to a state of complete repose by her swaying body, whispering voice and soothing rustle of her lightly starched, spotlessly white apron, secured at the back of her ample waist by a large bow, the ends of which reached with her skirt to within an inch or two of the ground.

My mother was born within the sound of Bow Bells and spent the first fifteen years of her life in that part of London. Her father, John Greenip, was a Keswick man and as a young man made the long journey south to the 'big smoke' for the purpose of trying his luck in the drapery trade.

It is apparent from an old newspaper cutting that his departure from Cumberland was important enough locally to warrant, or perhaps provide an excuse, for a farewell party which was attended by a goodly number of well-wishers and held at The Queens Arms, one of Keswicks' prominent hostelries of the day. Short of someone leaving his or her home town to spend the rest of their lives on the moon, I doubt if anyone's departure from anywhere in these days could excite little more than a wave of the hand.

My maternal grandfather's search for fame started, although I know not in what capacity, at the Bon Marche, Kilburn where he met and married Maria Selfe who was the staff housekeeper, catering for those members of the firm who lived-in.

Sad to relate John Greenip died at the early age of 42, when my mother Kate, or Kitty as she was more often called, being the middle one of three girls born of the marriage, was in her early teens. Destiny thus decreed that I should never see or know Grampy Greenip. There have been times when I have regretted not being able to fill in adequately in my own mind this partial void in my fairly recent ancestry, especially as he was the only grandparent I didn't know. Mother did of course leave me with a few impressions, but alas her family life with him was short. One thing was for certain, Granny being left a widow, with not very much money and three small daughters to bring up in the 1890's posed some almighty problems. However, it is said 'There is no strength without struggle', she managed somehow and I can only assume that courage, determination and hard work had a great deal to do with the outcome.

While in Keswick a few years ago I made a brief effort to find out something of Grampy Greenip's family line. A visit to the churchyard revealed on one tombstone the name of Greenip — while others displayed Greenup and Greenop who seemed to be rather more prolific. Maybe over the years the I's and O's and U's have been carelessly or purposely interchanged and they are all more or less the same family. However, as Granny spelt her name with an I, I came away feeling reasonably certain that I had seen the grave of my great grandfather Will Greenip — Born August 24th, 1819. Died November 2nd, 1895. Furthermore, removal of a fair amount of lichen from the headstone, by use of a penknife revealed not only that he was a naturalist and rural postman but also the following verse:—

> **In simple life he humbly sought**
> **To know the works that Nature wrought**
> **And none too poor to win his love**

If my assumption of our relationship is correct, great grandfather was a countryman indeed and one of his great grandsons at least returned to the fold via various circumstances some years later.

Gazing at the scenery around Keswick, with its craggy fells rising high into the sky, dotted as they are during the summer with specks of greyish white, which are the hardy breeds of sheep kept by those equally hardy farmers with whom he obviously came into contact, I don't find it difficult to visualise what sort of a man he probably was. It would seem that he enriched his work by studying nature at the same time. While an outdoor life of this kind provides this kind of opportunity I cannot help thinking that it is possible many of us caught up in today's rat race, fail to take even the slightest chance to comprehend what is going on around us, not only in nature but in the intimate sphere of human life itself.

Maria Selfe who was to become my maternal grandmother was born in the village of Shellingford about three miles from Faringdon. I believe her to have been one of a family of three boys and three girls. Her brother Tom I remember well, he died in retirement at Chirton in Wiltshire, after a life spent as an agent on one or two large estates. While I was a pupil at Dauntseys School in the late 1920's I was sometimes taken during term time to his house for tea. He was a tall, very upright man, and would have been very much in the fashion today as he wore his ample crop of almost white hair in a kind of 'bob' with a distinct parting from front to back, and it always seemed to stay in place. He mostly attired himself in a knicker-bocker suit with a belted jacket, which seemed to match his past calling and great love of shooting. He always drank his tea from a huge tea-cup, similar in size to a child's enamel potty, maintaining it tasted better in bulk so to speak, but I always thought it must have been well nigh cold by the time he had finished it — perhaps this added to the flavour. I remember too he kept a considerable cellar of home-made wines of which, as a treat, I was sometimes given a taste before returning to school.

I recall Great Uncle Tom reciting an interesting story concerning Silbury Hill, a huge cone shaped earth work close to Avebury by the side of the Old Bath Road, now the A4. The hill has fascinated archaeologists and others for many years, without I think, revealing much evidence as to its origin. I suppose one needs to have seen it to appreciate fully Uncle Tom's story, which was that he bet a friend a golden sovereign that he could run up to the top of the hill and down again before his friend could run the whole way round the perimeter at the bottom. The outcome of this somewhat unique gamble being that Uncle Tom became a sovereign better off. It is an interesting thought that the present day value of the coin at stake is currently around £65 to £70. He also took a great pride in his garden and well stocked orchard, and I remember him recalling an incident about the latter.

Apparently while out in his garden one evening he spotted two village boys making their way through the orchard hedge with obvious intent. In order to cause minimum disturbance and to remain for the most part out of sight, they climbed the tree of their choice, oblivious of the fact that their little scheme had already been observed. As soon as they were well up into the branches great Uncle Tom popped quickly into the house, collected his double-barrel shot gun and a couple of cartridges. He then sauntered casually, gun over his shoulder, to a spot immediately underneath the 'occupied' tree, where he stood in silence, without the slightest suspicion of an upward glance. After quite a while he took the two cartridges from his pocket, broke his gun for loading, uttering aloud, as if to himself, 'Ah well, don't suppose the birds will be back for any fruit this evening, but better let off a couple of barrels into the branches just to make sure'. In a split second the desired effect was achieved and frightened shouts from above indicated that it might be a long time before great Uncle Tom's orchard was raided again.

The product of any orchard in those days was looked upon as fair game. Within reasonable limits, 'pinching' was not really 'stealing'! Although there was the occasional professional job, when a small orchard was stripped one night and a message on the gate read simply 'All is safely gathered in'! Which is on a par with the poultry thieves who chalked this somewhat Robin Hood type verse on the door of the empty hen-roost a few days before Christmas:

> **Have robbed the rich**
> **To feed the poor**
> **Left cock and hen**
> **To breed some more.**

Up to about 1930 most farms possessed a small well cared

for orchard of apple and pear trees, the fruits of which were sometimes catalogued for sale by local auctioneers and sold at annual Orchard Sales, the purchasers doing the picking. I remember my father having more than one little gamble with the late Phil Wentworth, when he lived at Uffington. In early May, Phil would come and cast his eye over our old orchard, when the trees were in full blossom. He and father would then enjoy spending a while leaning on the old orchard gate haggling over the price, which didn't involve the winning or losing of a fortune for either but had the added ingredient of chance in which the birds and bees, as well as the wind and possible late appearance of Jack Frost, meant success or failure for one of them some four months later!!

The Wentworths were a large kindly family coming to this neighbourhood from Aldbourne in Wiltshire, as I believe some of my paternal forebears did. They were expert makers of home-made wines from many varieties of fruits and vegetables. I well remember when playing football for Uffington (the Amber & Blacks!) in the early 1930's, being persuaded by team mates Harry and Norman Wentworth to take a few "slurps" from their bottle before we kicked-off; these were repeated at half time and if it was a cold, wet afternoon another little "tot" to keep us warm until we got home — there were no clubhouses, baths or showers in those days; any changing that had to be done was accomplished in the lea of a hedge, the side chosen, depending on which way the wind was blowing or if there were any gals about.

I have knowledge that the Wentworth brew, referred to by some locals as "Tanglefoot" for obvious reasons also caused problems for those wonderful comedy actors Tom Walls and Ralph Lynne. They were staying the weekend with John Betjeman who was then a film critic living at Garrads Farmhouse, Uffington and he took them round to see the Wentworth family and taste their wine — with the result that Mrs Betjeman had to break off cooking Sunday lunch, to fetch them home some while later in her pony-trap, as the 200 yard return journey by foot seemed to be beyond any of them.

There is another incident concerning the Wentworths which happened about the same time. They owned a small two acre field in Uffington adjoining Woolstone corner and the chief spokesman of a band of travelling gypsies, supported by 3 or 4 swarthy looking relatives, persuaded dear old Mr Wentworth senior to allow them into his field with their 6 or 7 horse-drawn caravans, clinching the arrangement by managing to press 2/6 in the palm of Mr Wentworth's hand and making it stick. It became apparent a few days later that the gypsies, when asked to move on, considered the transaction which had taken place to be some sort of open-ended tenancy agreement and refused to leave. The drop in egg production in and around the village and in some cases the mysterious disappearance of hens and cockerels, as well as the drop in the wild rabbit population and the severe pruning of withy trees followed by intensive marketing of cloths pegs in the district, made it imperative that steps should somehow be taken to evict Mr Wentworth's so called tenants. The police were consulted but said they could do nothing as money had changed hands and they could only move them on once they were on the public highway.

So, with that in mind, one Sunday morning a posse of villagers and others recruited by the Wentworth boys entered the field by force, whereupon pandemonium broke lose, the gypsies loosed their horses and lurchers, shouting an hollering to set 'em alight, while the locals tried to push and pull the caravans out of the field on to the road. Whether the effort was ill-timed I know not, but it was Sunday midday and while local chicken may or may not have been on the gypsy menu, potatoes certainly were and the sight of saucepans of boiling spuds coming at 'em from caravan doors and windows had to be seen to be believed. I think the only casualty on our side was poor old Fred Ridley who lived in a cottage adjoining the field, and kept a considerable number of hens and a few goats, which he swore were virtually dry when he went to milk them every morning while the gypsies were in occupation. He actually received a direct hit on his sparsely covered head from a fair helping of gravy which, fortunately had just gone off the boil, when it was delivered through a slit in the canvas at the rear of a caravan. However, the mission was accomplished, not without a certain amount of undesirable language being used by both sides — the police moved them on as they said they would — to Longcot I believe.

It must have been around this time also that Phil Wentworth, the second eldest son started on the road which lead him to set up one, if not, I believe, the first supermarket in the district at Stanford in the Vale, where it is owned and managed by his son, John, today.

I recall Phil's first venture as being in the transport business — it was when all the milk within about 3 miles or so of Uffington Station was put on rail there in 17 gallon churns consigned to the Express Dairy Co, London, who had recently purchased the Buscot & North Berks Dairy business at Faringdon. The milk was taken to the station at about 7.30 every morning by horse drawn milk-carts of various designs most of them capable of carrying 6 to 8 churns. This was quite a chore every morning and there were difficulties during the winter when roads were icy and the horses were unable to keep their feet and needed to be frost nailed before they could be put on the road.

Frost nailing in itself was a fiddling kind of job, but with a train to catch, tempers of both man and horse could get a bit stretched. The aim was to insert about four pointed metal studs into each shoe by screwing them into holes already tapped for the purpose. The holes were cleaned out with a small tapping key to renew the screw threads to enable the studs to be screwed in securely by the use of a small box type spanner. The other form of frost nailing was much simpler and meant the removal of about two ordinary nails from each shoe and replacing them with bigger headed nails but to do this one had to be a bit of a blacksmith and be very careful not to prick the horse's foot and cause it to go lame.

Obviously, Phil had the notion there might be a demand for someone to transport the milk from farm to railway station at a reasonable charge so he purchased an old Ford Model 'T' motor car, knocked the back off and fitted a platform to the chassis capable of carrying about 10 churns, which were secured by a chain to the main body of the unit to prevent them slipping off the sides or rear. I presume there were very few restrictions and regulations to comply with at that time, certainly no M.O.T. The enterprise was supported in the first instance by the Whitfield Bros. of Fawler Farm and no doubt provided something towards Phil's next venture which was to knock a hole in the wall of an out-house adjoining their old farmhouse, alongside the road in Uffington, for the purpose of selling meat. It could hardly be called a butcher's shop initially, but it saved villagers having to visit town. Proving successful, the addition of a delivery van enabled a much wider area to be covered and he established himself as a master butcher at the time of getting married, buying property in Stanford in the Vale, converting it into a real butcher's shop with its own small slaughterhouse and from this developed the comprehensive supermarket that exists on the site today.

I feel sure Phil would not have minded me recording that around the 1930's if anyone playing darts in a pub around here had the misfortune to hit the wire of the double required and the dart finished up on the floor, many a wag has been heard to shout "Wentworth's Meat" (another way of saying "Tough"!). Similarly, if someone scored 26, the scorer was informed "Britchcombe" — 2/6 being the price in those days for a nights Bed & Breakfast advertised by the late Mrs Ayres on a board outside Britchcombe Farm, close to White Horse Hill where her daughter Marcella now farms and her granddaughter Ruth also runs a well patronised cream tea and catering business.

Granny Ayres, an aunt of the well known poet and T.V. personality, Pam Ayres, was a fine strong lady, unafraid to work beside any man and in fact did so regularly. As the Gaffer's wife she was, of course, entitled to set the pace so to speak, but on one occasion I'm told her enthusiasm led to having to take time off during root harvest, following the despatch by an employee of a fair sized mangel with rather more force than was required to get it into the cart from the opposite side to which Granny Ayres was working. Grampy Ayres was a keen grower of sainfoin which he made into hay and I think sold to the racehorse trainers around the Lambourn area. It served as an excellent break crop, grew well on the chalk and had the added attraction of adorning some of the land close to The White Horse with a beautiful crimson hue when it was in full flower.

Mention of Pam Ayres reminds me that her mother, whose maiden name was Phyllis Loder, was born in Baulking and attended the village school at the same time as I did. Pretty girl she was too.

To anyone who has been patient enough to read this far, I apologise for the lack of order in which I am trying to record what is in my mind. I think it is because I am scribbling very much on impulse and already I am aware that one thing or person mentioned leads quickly to others which makes a nonsense of any desire I may have started with to present a strictly chronological order of things, so please forgive the deviations and bear with me if you think it's worth it. I shall endeavour to return perhaps abruptly at times to an order based roughly on the succession of my birthdays!

Having previously mentioned great uncle Tom Selfe, I am reminded he had a brother Sam, who I never knew but was told more than once was a bit of a character and apt to gaze upon the vintage when t'was crimson, in short, enjoyed his tipple. I must be honest and record that a few contemporaries who outlived him, gave me to understand that in some respects I resembled him — they didn't indicate which. My father did in fact often call me Sam, so perhaps there was something in it. Be that as it may, Sam Selfe was a butcher by trade, carrying out his business in the Cornmarket, Faringdon on premises occupied later in succession for the same purpose by two well known Faringdon families, the Taylors and the Carters. The site housed one, if not the last, licensed slaughter house in the district, the approach to which being under a large stone archway, was the original main gateway to Faringdon House many years ago. Recent change of ownership means the ladies dresses now hang from rails where large hooks once supported whole sides of meat.

Sam remained a bachelor and I believe lived at Chapel House, Great Coxwell from where he often sallied forth in a high trap, drawn by an equally high spirited but very faithful steed which on numerous occasions knew the way home far better than Sam. Once home this equine wonder would stand in the yard and wait for his master's housekeeper, Miss Rixon, to take him out of the shafts which were dropped to the ground for the express purpose of removing great Uncle Sam from the trap! While such an exit may have been somewhat undignified for Sam, Miss Rixon had come to learn that no other method achieved such immediate success.

It is amazing what homing instincts many horses possess. I have heard it said that before motor cars came on the roads, a local farmer's supper and card party broke up in the early hours of the morning and the host having previously instructed his groom/handyman — who had stayed up to harness horses and see the well inebriated guests on their individual ways home — to put the horses into traps which were different from the ones in which they arrived. The story as told to me did not conclude by saying how many failed to recognise their own bedroom doors, if they ever got that far. The only tangible reminder I have of Uncle Sam is a small pair of silver sugar tongs handed down to me with the story that they became his property as a result of one of his customers being continually unable to find sufficient 'ready' with which to pay for the privilege of having consumed prime cuts of Uncle Sam's "Roast Beef of Old England".

Of my maternal grandmother's two sisters one married someone by the name of Cave and lived in Uxbridge area and although I recall my mother talking of her cousin Maud Cave I fear somehow that branch has withered. Her other sister I do remember, she married a John Bury. They lived and farmed a small farm at Cherrington in Warwickshire where I was occasionally taken for a Sunday visit in my childhood days.

It was here that I became fascinated with a cream separator made by Listers of Dursley which I was allowed to turn by hand at what seemed like ever increasing speed, the faster it went the higher the whining noise it made as the milk trickled from a three gallon receiver down into a centrifuge which, revolving at speed, separated the cream from the milk which dripped out of one spout while the skimmed milk ran out of another. The result of the exercise being that butter was made for sale locally, the skim being added to wheat toppings and barley meal and fed to the pigs. Another thing I remember about visiting Cherrington was that aided by second cousin Betty Bury I was able to play "Now the day is over" with one finger on their piano. Whether this encouraged my mother to think I possessed some talent in this respect I know not, but I shudder to think of the money spent at home and school on trying to teach me to play the piano and am sure it could have been better spent.

As I mentioned earlier Granny Greenip née Selfe was born at Shellingford in a house now named Hollywell House which stands opposite the old forge on the T junction leading down the "no through road" towards the church, school and the main part of the village.

It would seem that upon the death of her father Robert Selfe in 1891 whom I presume she had been looking after, her mother having died previously, she left to take up a house keeping post in London, where as I have earlier mentioned she met and married John Greennip. Before she left she composed the following poem of which I have a copy in my mothers handwriting, it is entitled:

On Leaving Shellingford in 1891

Goodbye to the old pleasant lawn
Where we read, served or talked while at tea
In the dear summer days of our childhood
When the birds sang merrily.

How nice in the cold days of winter
On a clear and bright healthy morn
When the only little excitement
Was the sound of the huntsman's horn.

We could hear the hounds all barking
In field or wood close by
Poor Reynard rushing wildly
With horse and rider in full cry.

Goodbye to the neighbours next door
And the dogs, both Terry and Nell
The old blacksmith's shop in the distance
With the sound of its iron anvil.

Farewell to the quiet sleepy street
And honest village people
The quaint old fashioned church
Its belfry and its steeple.

Goodbye to the green church yard
Where Father and Mother lay
Awaiting the final summons
To everlasting day.

Until I am laid beside them
Whenever that call may be
I shall always have fond recollections
O thee my once happy home.

M.M.Selfe

Until a few years ago Shellingford boasted a very unusual licensed drinking place. It wasn't exactly a pub, it was a sort of Post Office cum shop, cum off-licence with no indoor bar but a "hole in the wall" through which one could sing out for a pint, pay for it and drink it standing in the village street. It was known locally as the "Ole in the Wall" it's last proprietor being I think a Mr Poole who also ran a village taxi service. I imagine this unique opportunity to booze could be very pleasant during the summer and most convenient for haymakers and harvesters on their way home at the end of the day but I doubt it attracted more than two or three hardy villagers in the winter. At least there was no "chucking out" to do!

My knowledge of the Selfe family's life at Shellingford is very sketchy and incomplete I only wish I had asked a lot more questions before, alas, it was too late. I do know however that one branch of the family farmed at Little Newbury Farm, maybe for a Mr Calvert, before the Kitemore Estate was split up and sold in 1923. I have previously mentioned my maternal Grandmother's two brothers, Tom and Sam. Granny also had another brother Philip who once farmed at Wickwood Farm, Shellingford. As far as I know he had three sons and a daughter. One son Herbert had a furniture and upholstery business in the Market Place, Faringdon; it was a double fronted shop, next to the present Social Club and is now divided into a flower shop and delicatessen, incorporating a small cafe. The two other sons, Gerald and Cecil, emigrated to Africa to set up a farming enterprise near Bulawayo in Southern Rhodesia where they were joined later by a Scotsman named Jock Brebner who had recently married their sister Ethel.

Apparently Cecil had been quite fond of his cousin Kate (my mother) before emigrating, and when he returned to England for a short while, having been picked to represent the Rhodesian Forces at King George V's coronation, old flames were rekindled and he asked her to marry him and accompany him back to Rhodesia. I remember my mother telling me this one day when she was in a reminiscent mood and how she had to decide whether to accept her cousin's proposal or stay in England and marry father, who had been courting her for sometime. It will be seen from the reproduction of the original letter received by Cecil's family, how very sad the outcome would have been should the coin have fallen on the other side; strange what fate has in store and how the cookie crumbles. It is just possible I might have been in Zimbabwe at this moment instead of sitting here at Baulking writing this.

Selfe— On Dec. 1st, of Dysentry whilst serving with the Southern Rhodesian Volunteers, Cecil E. youngest son of the late Philip Selfe, of Wickwood Farm, Shellingford, Faringdon, aged 37 years.

The Casualty List.

News has just.... been received by the relatives of the death of Cpl:- Cecil E. Selfe, who succumbed to Dysentry on Dec 1st, whilst serving with the Southern Rhodesian Volunteers on the Northern Borders of Rhodesia. The name of the place is not given, but some idea of the nature of the country in which these troops were may be gathered from a recent letter in which his sister states that her brothers were 60 miles from the base, the only means of communication was by native runners. Cpl:- Selfe who was a man of exceptionally fine physique, was the youngest son of the late Philip Selfe, of Wickwood Farm, Shellingford. He served through the South African War as a Sergeant in the Wilts. Yeomanry, was mentioned in despatches. Shortly after the conclusion of hostilities he took up farming near Bulawayo, where he was enjoying a most prosperous career, in which he had been recently joined by his brother-in-law. He was one of the picked men to represent the Rhodesian Forces at the King's Coronation and this was the occasion of his last visit to his native land.

At the outbreak of the present war he again volunteered for the front, and at the time of his death was serving with his brother, Gerald, in the above-mentioned regiment. He was 37 years of age.

Turning to my paternal grandparents I have always understood that the Liddiards moved into the Faringdon area of Berkshire from Aldbourne in Wiltshire but exactly when I know not. I have a copy of the family tree starting with John Liddiard born 1736, descending through seven generations to my youngest grandson James born in 1977.

I am grateful to whoever compiled the earlier information therein but it would have been even more helpful and interesting if the locations of the "arrivals" and "departures" had been recorded as well. No doubt research would complete the picture but perhaps that is for some one else another day.

I just remember my Great Grandfather George who died aged 90 in 1918, and I appear in a family photograph *(see frontispiece, p.4)* representing the youngest of four generations of Liddiards, he being the eldest. It was taken on the lawn at Coxwell Villa in 1914 by Tom Reevley, a well known photographer who ran a studio in The Market Place, Wantage. There must be a large amount of his work still around in many homes in this area today and I can recall seeing him disappear under the black cloth covering the rear of his camera on more than one occasion, pressing some sort of button on the end of a cable for a more instant operation.

My grandfather was George Ernest, known best to everyone as Ernest. To me he has always been a character that has grown with the years and there is quite a lot I would like to say about him. He was a smallish man, weighing about 10 stone, who attired himself mainly in tailor made suits, or riding breeches and cloth gaiters, topped mostly by a black bowler hat which was replaced by one of a silver hue on high days and holidays. His suits were cut and made by Percy Pocock, a gentleman's tailor, so it said, in semi-circular lettering on the window of his shop in Southampton Street, Faringdon, which is now part of Barclays Bank premises. It is my one and only recollection of seeing a tailor sitting crossed-legged on a table doing his work, surrounded by cloth and tools of his trade. Mr Pocock himself was a tall, upright smartly dressed gentleman, a credit to his profession, wearing a gold watch and chain and carrying a silver-topped walking stick. So attired, he was asked the time by a customer whom he chanced to meet while on his evening stroll along Gravel Walk, giving him the welcome opportunity to reply "It's nearly time you paid me for that last suit I made for you"!! "Really," the customer replied, "I didn't think it was so late by five and twenty minutes"!

Grampy's coats were always cut away at the front to form swallow tails at the rear with a couple of buttons in the small of the back. The legs of his trousers were very tight fitting indeed. A lady once asked me if I knew how he got into them and I told her Granny stood at the bottom of the stairs every morning, holding them open, and Grampy ran down the stairs and jumped into them.

He was an extremely energetic man, with an obvious appetite for living, which entailed a great deal of hard work and a distinct flare for getting things organised so that he could retire from his grocery business in the Market Place at Faringdon in his early fifties and enjoy the fruits of his labour doing a bit of hobby farming at Steeds Farm, Great Coxwell, fox-hunting in the winter, and helping to run the Old Berks Point-to-Point of which he was Hon. Secretary for a number of years when it was held at Baulking, Lew, Middle Leaze, Coleshill and Step Farm, Faringdon. In the summer his activities were directed towards running the Faringdon Whit-Monday Sports on the old Sports Field where Faringdon Town played football in those days. An area which no longer echoes to the sound of leather upon leather and the shouts of a few local supporters but the screaming of rubber upon tarmac, as the far end set of goal posts must have stood almost in the middle of the new roundabout on the bye-pass at the eastern end of the town. Grampy also had a great deal to do with the North Berks Show which alternated between sites at Faringdon, Wantage and Abingdon before it became amalgamated with the Oxfordshire County Show which itself folded up during the early 1950's.

I am indebted to Chris France, a retired auctioneer and

formerly a partner in Moore, Allen and Innocent, the estate agents of Lechlade for telling me he had come across a story in an old "Faringdon Advertiser" a weekly newspaper dated the 4th of October 1913. It contained a detailed report of the North Berks Show held the previous week in the grounds of Faringdon House and was told by the Show's President, Sir Alex Henderson M.P. at a "capital" luncheon in which he thanked everybody concerned for their work and enthusiasm in making the event such a great success. Saying "Perhaps their thanks were particularly due to Mr Ernest Liddiard who had so much to do with the organisation, and possibly no small part of his duties had been collecting money for the local prize fund which this year was a record amount," (applause). Sir Alex went on to say that "he hoped he was not telling stories or facts that should not be disclosed but he had heard that a small boy in Faringdon had recently swallowed a silver sixpence and on arrival at the Cottage Hospital he kept crying out "Send for Mr Liddiard", whereupon he was told it was a Doctor he needed, not Mr Liddiard, but he persisted with his request and when asked why, he replied he had heard his father say a few days ago that Mr Liddiard could get money out of anyone. (Laughter and cheers).

His love of sport prompted him to donate a piece of land, being part of an orchard in Southampton Street for the purpose of setting up the Faringdon Bowling Club. I have a vague recollection this was immediately following the Armistice in 1918 as I recall my father telling me that German P.O.W's helped to lay the turf for the rinks, which came from Cumberland by train to Faringdon. Trundling the woods seemed to be enjoyed by quite a few of the Liddiards. Grampy obviously did, his brother Frank played for Berkshire, and father's name is on the Honour's board in the present Club-house in Canada Lane where the Club moved when the bye-pass threatened to swallow them up. It is an honour that the Club continues to appoint a Liddiard as one of their Trustees. My father, myself, and now my son Bill have participated as such and it would be nice to think it may continue.

Grampy also donated a further piece of land adjoining the original Bowling Club site to the Faringdon Lawn Tennis Club, who then moved from its old venue at the end of Church Path overlooking Faringdon Grove to its present location where it currently provides pleasurable sporting and social facilities for present-day Faringdonians which I am sure was the purpose of his generosity.

It wasn't until quite a while after he died and as I grew older, that I began to appreciate some of his qualities and tenacity of purpose, and realised I was being influenced to some extent by the twenty two years I had known him during which as a small child I loved him, mainly perhaps because I recited various children's rhymes such as The Jolly Miller and The Crooked Man Who Walked a Crooked Mile, etc. and he rewarded me with a shining sixpence and allowed me to kiss his shaven, almost bald pate! I don't think he was a particularly affectionate man — only Granny would have known that for certain. Anyway he seldom showed it and at times he could be a bit sharp which in my early teens frightened me. I remember him having one or two goes at my father about his farming efforts in the middle of two of the worst depressions there have ever been, during the 1920's and early 30's. At the time, I thought it was unfair and that father was doing his best despite his weakness for a bit of horse-dealing and the fact that he was really a grocer by trade, becoming a farmer by reason of having been in partnership with his older brother Stanley when the Great War broke out.

I felt it might be interesting to recall something of Grampy's business as a Wholesale/Retail Grocer and the attached subsidiaries of a bacon factory and mineral bottling plant. There are still a few old marble topped bottles around with Liddiard embossed on them, I am told, and Fortnum & Mason of London were valued customers especially at Christmas time for Collared Head, a brawn made to a particular recipe from pigs' heads. The head slaughterman's perks such as brains and bladders, the latter dried for me to play with, like balloons, occasionally found their way to mother's kitchen door for conversion into something a bit more drinkable.

That Grampy was energetic, is surely an understatement in relation to the fact that once a fortnight, when they lived over the shop in Faringdon Market Place, Granny would have given him breakfast, warmed his driving coat and bowler hat by 3.30am. A young lad would have a fast trotting cob in a trap ready outside to go with Grampy to Swindon Station where Grampy would catch the early workman's train to Bristol, a journey which took a few minutes over the hour, while the lad took the cob to stable in the Gt. Western Hotel opposite, kipping down until Grampy arrived back at Swindon; having done what he wanted to in Bristol by way of bulk purchases etc. He would then drive back to Faringdon in time to enjoy three or four hours hunting depending on how near the hounds were meeting that day.

It would be interesting to know if there would be anyone in Bristol today prepared to do business at such an early hour, let alone a suitable railway time-table to make it possible. Perhaps there is — anyway I suppose modern cars and the M4 would turn a similar exercise into a "doddle" these days, and you probably wouldn't need to sit down to breakfast before 7am and still have your day's hunting.

Another instance, Grampy broke his leg out hunting, which meant huge wooden splints being bandaged in place to allow the break to heal and a week or two in bed. As my father and his brother Stanley were only in their early teens, Grampy decided they were not quite old enough to take charge of the shop. So he had a small trap door cut in the bedroom floor above the shop counter below so that he could give orders to the staff or speak to customers if necessary. It is interesting that despite several changes of ownership and many alterations to the property, the evidence of this method of communication is still very plain to see.

The wholesale side of the business necessitated a fair sized stable of horses, consisting of a few trappers but mostly vanners which were a type somewhere between a cob and a carthorse. These were used to draw the large wooden wheeled, iron tyred grocery vans heavily laden with goods to a large number of village shops within a 15 mile or so radius of Faringdon, some of which were assisted financially and offered various terms, such as discounts, deferred payments etc. When deliveries were made to villages such as Lambourn, for instance, arrangements existed whereby the vans were met at the bottom of Ashbury Hill by a farm carter with a team of trace horses to assist them to the top of the hill, before uncoupling and returning to the farmer who provided this service and for which, of course, he was financially rewarded or perhaps given a bottle of port.

Mention of which reminds me of a story told to me by my father. Apparently, it was part of the business to buy large wooden casks of port which were decanted and sold off in bottles.

Grampy noticed that the number of bottles being decanted per cask seemed to be on the decline, so in order to find out where the shortfall was going he told my father to visit Ballards the Chemist to obtain a fairly strong mixture of what he called "JOLLOP" and, when the next cask being decanted was almost empty, to pour this in. He also told him at the same time to sprinkle the box of paper in the staff loo with a liberal amount of black pepper adding that he didn't think it would be long before the culprit came to light — and it wasn't.

Grampy was a very hospitable man — perhaps that particular member of his staff might have thought, too hospitable. Market Days were on Tuesday, and the first Tuesday in each month was known as Great Market. It was always open house on such days and many farmers' friends and tradesmen were invited in to hot lunch preceded usually by a tot or two of Gin & Ginger known variously as "Old Berkshires" and "Longcot Thorns" the latter, so named after Harry George who dispensed them liberally at his home, Cleveland Farm, Longcot, especially when hounds met there annually to help celebrate his birthday.

The first draw that day was always at Longcot Thorns, on Harry's farm, hence the connection. One thing was certain, partakers of this blend of stirrup cup were assured of being in the right frame of mind to enjoy the fun, the hedges and ditches appearing to be half their usual size and body temperature would remain at a comfortable level for quite some time despite what the elements chose to throw in.

The original mixture of the drink was half a bottle of Gilbey's Gin mixed with half a bottle of Schweppes ginger wine. I don't think Schweppes make a ginger wine these days but Stones or Crabbies, depending on taste, will do. Harry George always maintained it was better when mixed at least 24 hours before consumption — he said it kind of mellowed a bit — he should know, he tasted enough of it. A warning though; it's terrible stuff to get drunk on; ashamedly I speak from experience. I thought I was going to die and at the time I almost wished I could. So to the adventurous, beware, it's a real party livener and tongue loosener as some of my friends will know especially perhaps John Pollard, a past Principal of The Berkshire College of Agriculture who admits to making the speech of his life, and he made many, following a little get-together at Church Farm prior to a Faringdon N.F.U. Dinner at which as Chairman, I had invited him to hold forth as Guest Speaker. I remember him saying afterwards the drink needed to be renamed "An Old Berkshire Bomb"!! Forgive me John. I hope you enjoyed the drink as much as we enjoyed your speech.

I was introduced to "Gin and Ginger" by Granny Liddiard in sensible easy stages when I was about 10 years old and they lived at Romsey House, Gloucester Street, Faringdon.

Stage 1. consisted of a small drop of ginger wine diluted in water which was followed a couple of years or so later with neat Ginger Wine to which as time went by gin was added in increasing amounts until Granny was satisfied I was old enough to cope with a sherry glass full of the real stuff.

Harry George and his wife Annie, who was a dear soul, were great friends of my mother and father. Before settling down to farm at Cleveland Farm Longcot, they were Licensees of "The Stewart Arms" a London public house close to the White City at the time when the Great Exhibition was staged there during the turn of the century. I recall Harry saying more than once that although the "drink" nearly killed him he made quite a bit of money to enable him to farm and enjoy hunting with the Old Berks. Harry was quite a character, looking every inch a jovial rosy faced landlord. Quite prepared to employ others to do the manual part of farming while he looked after the business and social side of life.

When in London, he bought himself an early model Ford two seater motor car complete with brass fittings, bucket seats and a vertical steering column mounted, of course, by the steering wheel underneath which was the hand throttle. Gear change was by a combined clutch and foot pedal and the hand brake was situated on the right hand side of the driver on the outside of the vehicle. Quite an impressive machine it must have been. The only comparable model I can remember seeing in action belonged to a Mr Ferryman, the estate agent to Viscount Barrington who lived at Becket House, Shrivenham. I also remember someone slipping a huge wooden log under one of the rear wheels of his car, when Mr Ferryman called for some petrol at Brickell's Garage on the outskirts of Shrivenham. After numerous unsuccessful efforts to drive off, he told the petrol pump attendant he could walk to his home in the village as he thought the brakes had seized up, and would he ask Mr Brickell to deal with the trouble and return the car to his home, which he was more than pleased to do accompanied by a small charge of 10 shillings for releasing the braking system; incidentally Brickells' Garage was the first garage I saw using hand operated petrol pumps.

I digress. One of Harry George's first journeys was to take his wife from London down to spend the week-end at Longcot. Confessing to knowing nothing at all about the mechanics of motor cars nor wishing to, he set off on the sixty mile journey. Managing to get out of London without too much trouble, apart from getting stuck in one of two tram-lines, his story relates that having descended the hill into Henley-on-Thames he was faced with extensive road works taking place on the bridge over the river and the scene was littered with barrels and scaffold poles etc. Harry said before he could think what to do he was in the middle of it all. The commotion soon brought a crowd of on-lookers, some of which were voicing the opinion that these b....... motor cars should not be allowed on the roads at all. The arrival of a policeman on his bicycle and his immediate suggestion to Harry that it would be best if he and his motor car left as quickly as possible, prompted Harry to ask him to turn the handle in the front to see if the engine would start. It met with immediate success and as the car was still in gear it leaped forward knocking the policeman to the ground who, Harry said, he never saw again as he sped through the rest of the town out into the countryside towards Longcot. But his troubles were not over, as approaching Cleveland Farm he was unable to remember how to bring the car to a halt, but his knowledge of the layout of the farmyard, and the fact that the farm gate on the Watchfield Road was open enabled him to drive in and bring the car to rest with its bonnet partly buried in the manure heap close to the main cattle shed.

Harry never really mastered the art of driving a motor car. When he left London and came to live at Cleveland Farm he had a privet hedge planted by the side of the driveway leading from the garage so that when reversing out, he could keep an eye on the hedge without turning his head or having to use the mirror.

Harry loved his hunting though, not that he was a fearless rider across country but he enjoyed life and people. He was always well turned out by his groom Leonard and usually wore a small bunch of violets or something similar in his button-hole. On one occasion while hacking to a meet of hounds at Broad Hinton he passed the tenant of Ruffian's Wick Farm, a Mr West, who was busy cutting and laying a hedge by the side of the road. Following a cheery "Good Morning" and a brief chat, Harry said, "You should be on a horse and coming along with me to enjoy yourself," to which Mr West, standing there wearing an old hired sack tied around his waist to keep the thorns from pricking his knees and a pair of thick leather hedging gloves, clutching a billhook in his right hand, replied, "Bugger, Farmer George, I be enjoying myself". I can only comment that such a remark confirms the old adage that "It takes all sorts to make a world".

On Market Days, Faringdon Market Place was full of livestock which were auctioned by Messrs Hobbs & Chambers. The dairy cows and beef cattle were penned in what were called cow-flake hurdles, (they were bigger than sheep hurdles), alongside the churchyard wall in Church Street; the auctioneer selling from inside a small hut on iron wheels which was trundled away at the conclusion of the day's business, as were the hurdles and other market trappings before the area was swilled down from a water barrel or hosepipe when it became available later. Calves were tied to a rail in front of the now Social Club. Sheep were hurdled in front of Grampy's old shop next to Lloyds Bank which, towards the end of the 18th Century, was the City of Gloucester Bank. The pigs were penned in front of the Old Town Hall almost in front of The Crown.

There were of course very few motor cars around in those days but as they increased in number, traffic on market days became a problem as did the "residue" left behind by livestock. I'm not quite sure where the old saying "Where there's muck there's money" comes in here but there must be some sort of relevance. However, in the early 30's the auctioneers, in consultation with the Faringdon Branch of the National Farmer's Union and others, moved the selling of Dairy and Beef cattle to a site on the corner of Coach Lane and Church Path and equipped the area with permanent pens and cattle weighing machine at the entrance to the sale ring. The selling of calves, sheep and pigs remained in the Market Place until just after the Second World War when it was transferred to the Coach Lane site in the late 40's.

Eventually due to the swift development of motorised transport in the late 20's and early 30's and the arrival of cattle lorries which were capable of transporting three or four 12 cwt beasts at a time to larger markets further afield such as Banbury, Reading, Gloucester and later on Chippenham, the smaller town cattle markets gradually disappeared. I think it is fair to say this process was speeded up where auctioneers became more involved in the developing property market and less reliant on the farming side of their business. Farm Sales around Michaelmas and Lady Day for instance were two a penny, fifty or sixty years ago, the bill posting boards in towns were plastered with them. The first cattle transporter I can remember in this area was Frank Cahill who started business at Little Coxwell. There was also Alf Moffatt and Tommy Clare, who founded the well known Eagle Motors Coach and removal van enterprise. I don't think he was ever involved in cattle transport — perhaps horses; he loved his bit of hunting. Its just occurred to me that he and Frank Cahill both worked for G.W.R. at Faringdon Station; may be their close connection with the transport of goods and people triggered off a desire to set up on their own in this particular field.

Faringdon between the wars, like other small market towns, had a fair number of shops catering for almost every need of its inhabitants and those of the surrounding district. There were at least four grocers, three butchers, two ironmongers, an agricultural engineer, four drapers, a shoe shop, two cycle shops, two saddlers, a corn merchant, three or four builders, two tailors, three bakers, a fishmonger, two hairdressers, a toy shop, two newsagents, a chemist, wine and spirit merchant, jewellers, dairyman and various small shops selling sweets and tobacco. There were also two Doctors, two Veterinary Surgeons, the Cottage Hospital, which is now The Health Centre and about twenty pubs and hotels. Larger businesses were the Buscot & North Berks Dairy later taken over by Express Dairies and a Saw Mill, both being close to Faringdon Railway Station which was connected to the main G.W.R. Paddington to Bristol line at Uffington Junction about four miles distant. There were also two privately owned companies producing light and power for the town. The Gas Works was situated in Canada Lane which originally provided the town lighting and I recall seeing a shortish man with a long-handle hook which he inserted in the rings on two short chains just under the lamp itself to turn the gas on and off which was on seven or eight feet high lamp posts— I think the mantles ignited from a very small pilot flame which burned continuously. I suppose the lamp lighter's hours of work varied with the rising and setting of the sun. The gas works also provided a very useful by-product known as "gas tar" which it sold off in drums to local farmers for painting and preserving almost anything but particularly corrugated iron roofing, it needed a great deal of brushing on, but had a lovely healthy smell!! The Electricity Company's power station was in Southampton Street and stood on a site now serving as a car park for Barclays Bank. It was possible to stand in the street and watch through a large iron girded window what, to me as a small boy, was a huge beautifully maintained steam engine with a colossal fly wheel, powering the turbine which produced the electricity for the banks of accumulators used to store and dispense its output.

About half a dozen of the town's business owners, of whom Grampy was one, met regularly in one anothers houses during evenings for the purpose of a drink or two a bit of supper and a gossip about local affairs. For some reason Grampy liked having various pets around. I remember him having a small monkey for a while and a fox cub which became quite tame and, Granny said, stank like a pole-cat. He also kept various birds such as magpies and jackdaws.

On one of these summer evening get-togethers on the lawn at Romney House, to where Grampy moved upon retirement, a jackdaw was allowed freedom to roam the garden and join the proceedings with his squeaky calls of Jark — Jark. Like all jackdaws he was a particularly inquisitive bird and, unbeknown to those present, had discovered a glass of whisky on a tray on the lawn beside one of the guests, which he had obviously partaken of somewhat greedily as he proceeded to entertain everyone present before he finally passed out, fell over and was restored to his cage to sleep it off. Apparently he was fine next day apart from being a little quiet. Any way he was allowed to join in on similar occasions with the same kind of result; obviously some birds never learn!!

There is no doubt in my mind that the town politics were discussed during the hospitable gatherings just mentioned and it is possible interest and a certain amount of excitement was generated around election time for seats on the Parish Council. The notes and comments for one of these held in 1907 are reproduced below and it is quite possible I think that they were issued as a kind of supplement attached to the weekly newspaper printed in the town by S.W. Luker and named The Faringdon Advertiser.

They were handed down to me via my father and grandfather, and great grandfather who is actually on the list of runners as "Father of the Council":

From our Note Book.

THE PARISH COUNCIL STEEPLECHASE,
Run Monday, March 25th 1907.
****

In the usual order of things we should have issued a list of competitors in this important triennial event, a week or more ago, but as in this year's Grand National, matters have been peculiar, and several candidates have been and still are, under grave suspicion, that we have till now delayed in submitting to our patrons, that we trust will prove a correct list. However, the matter cannot be postponed longer so here they are:-

"Stop Watch" by "Timekeeper"—"My Jewel."
"The Gent" by "Wadley Hero"—"Royal Prize."
"Saddlers Pride" by "Stirrup Leader"—"Compo."
"Mixed Pickles" by "Commissioner"—"Doubtful Honor."
"Retirement" by "Home Cured"—"Industry."
"Foreman" by "Timber Bob"—"Sawdust."
"The Solicitor" by "Taxed Costs"—"Legal Quibble."

"Daily Bread" by "Finest Flower"—"Sweet Stuff."
"Flannelette" by "Discontent"—"Tape Measure"
"Justice of the Peace" by "Maltster"—"Popularity."
"Foggy Film" by "The Socialist"—"Work No More."
"Astley Don" by "Prime Cut"—"Tender Rump."
"Red Jersey" by "Salvation"—"Follow on."
"Father of the Council"—"Windfall."
"Prime Cheddar" by "Smoked Wilt"—"Choicest Blend."
"Bromsgrove Squire" by "County Councillor"—"Unopposed."
"New Hydea" by "Tailor Made"—"Sure Fit."
"Friendly James" by "Scholastic"—"Weather Profit."
"The Chairman by "Horrible Crime"—"Madame Tussaud."
"Confection" by "Tender Conscience"—"Jam Tart."
"I Would" by "Would I"—"I Wouldn't."
"Foiled Ambition" by "Dangerous Ground"—"Try No More."

The last two mentioned have, we hear, been scratched.

In the above list another strong resemblance to this year's Grand National, will at once strike you, viz:- There isn't a single mare running. It is a big entry nevertheless, and includes many well-known performers, that majority of whom have triumphed in previous contests, and in the meantime done honorable service to their supporters, but whether some of the untried and we may say unknown candidates, which are entered this year for the first time, will stand the severe test of training necessary for such an important event, is a matter of conjecture. In our humble opinion several of them will come such a severe cropper that they will be nowhere in the race, and will be returned a long way down in the list of "also ran".

We have taken great pains in weighing up the merits of all the candidates and have now no hesitation in advising our Patrons to plump solid for the following thirteen:- "Stop Watch," "The Gent," Saddler's Pride," "Retirement," "The Solicitor," "Daily Bread," "Justice of the Peace," "Astley Don," "Father of the Council," "Prime Cheddar," "Bromsgrove Squire," "New Hydea," and "Friendly James."

These well-trained competitiors with your support, are bound to win, for already there are signs of weakness in the opponent's camp. As before stated "I would" and "Foiled Ambitions" have completely broken down and have been scratched. Of the others, judging from recent form, "The

Chairman" is not the Horse he used to be, and "Mixed Pickles" is not to be trusted. "Flannelette," "Confection," have tried their luck in this contest before but failed, as they undoubtedly will this time. "Foreman" is practically unknown, whilst "Foggy Film" and "Red Jersey" may be passed over without comment.

Make no mistake. You have nothing to fear. Back the thirteen selected for all you are worth, and do yourselves some good.

One last word. Please don't forget your old friends and advisers when you are counting your winnings.

Yours obediently,
TWEEDLEDEE & TWEEDLEDUM.

The occupations of most of the contestants I came to know later (I wasn't born until 1913), were:

Abel	Jeweller, now Deacons
G. Adams	Farmer whose great grandsons still farm at Fernham
Burgess	Saddler, now Coopers
Carter	Grocer, now Supermarket
Chamberlain	Grocer, now Wine Merchant
Chapman	Not Known
Crowdy	Solicitor, now Crowdy and Rose
Fletcher	Baker London Street, now private house
Goddard	Draper; now Rats Castle Restaurant
Habgood	Brewer, now 2 shops
Hayworth	Photographer Gloucester Street, now private house
Heavens	Butcher, now Pat Thomas
Hunt	Not known
G. Liddiard	Grocer, now Tourist Centre
F. Liddiard	Grocer, now Bargain Centre
Lockwood	Photographer, now private house
Pocock	Gents Tailor, now part of Barclays Bank
—	
—	
Willis	Confectioner, now Marriotts Estate Agents

| Taylor | Butcher, now Morleys ladies wear |
| Proctor | Not known |

It is often said that behind every good man there is a good woman — so it was with Grampy — he married Mary Chillingworth in 1863. I remember her with a great deal of affection. She was a short, rather thick-set person, almost without a lap when she sat down in her own favourite armchair which seemed to be part of her and shook with her when she chuckled about something that amused her. I don't remember seeing her wearing anything other than black except for an occasional splash of white or grey on a silk scarf.

She was one of a large family of Chillingworths who farmed at Shellingford, before moving to Inkpen near Newbury from where the whole family emigrated to New Zealand, with the exception of Granny who was the oldest and the youngest son, Andrew, who was still at school.

It is possible the reason for Granny remaining was that there was a bit of courting going on involving Grampy as they both became unofficial guardians to Andrew, who when he left school assisted in Grampy's business for a while before going to help an Uncle Richard Chillingworth, who was in very poor health, and farmed at Queenlaines Farm, Sevenhampton, which was part of the Warnford Estate. Story has it that there was a somewhat passionate love affair going on between a daughter of the Warnford family and Uncle Richard but a union was prevented as it was unheard of that the daughter of a titled landowner should marry a tenant farmer. Further story has it that Uncle Richard altered his will in favour of Andrew for all the help he had been to him in running the farm, but his housekeeper got hold of the will and tore it up.

Of those who emigrated, I only have knowledge of two: Edie, who married a Mr Owen, and I recall listening to them talking about their life in New Zealand on one of their trips home to England, and the other, a brother named Joseph, who remained a bachelor, apparently doing quite well for himself as he amassed enough to retire early, live to the ripe old age of

90, quite a while of which was spent in a nursing home, died intestate, so leaving Granny's surviving children or their issue, a few dollars apiece — bless him! Upon the death of Richard Chillingworth, Granny's brother Andrew married a Miss Cole and was granted the tenancy of Queenlaines Farm. He was a good farmer and well known as a breeder and Judge of Dairy Shorthorns having officiated in this role at the Dairy Show. With his son Charlie he will also be remembered in the annals of Point-to-Point racing for having bred a wonderful mare named Prime Dutch who, ridden by Charlie, carried their colours, black, white stripes, white star on black cap to victory on so many occasions. She is the only horse I ever actually saw win two 3 ¼ mile races over fences in the same afternoon. It was at the V.W.H. Meeting at Blunsdon in 1923. She won the first race — The Farmer's Lightweight in which I think she carried 12 stone and the last on the card the Farmers Heavyweight in which she lumped round thirteen and a half stone and won easily. It was a great achievement for all concerned, as the mare was kept in trim between races by continual grooming with hay wisps by members of the farm staff, in between periods of walking exercise, to prevent stiffness and keep her muscles supple. I have some recollection of her being taken by train to contest The Lady Dudley Cup, the blue riband of Point-to-Pointing staged by the Worcestershire Hunt, the outcome being that the small posse of supporters who travelled in Uncle Andrew's "Essex" motor car failed to arrive in time for Charlie to walk the course beforehand. Consequently, having gone into the lead on the turn for home he somehow lost his way. Going very wide he lost a lot of ground and was squeezed out when joining the leaders at the third last fence. I think the race was won by a grey horse named O'Dell, owned and ridden by Major Rushton a great supporter of hunting and Point-to-Pointing in those days.

The only other horse I have heard of who won two races on the same day belonged to a Mr Woodin of Draycott Moor. His son Guy told me (he was only about six years old at the time) he thought it was a grey named Snowflake and was ridden by Billie Cooper of Manor Farm, Eaton, Nr Abingdon, where David Farrant now farms. Another horse who came very close to accomplishing the same feat was The Doctor, owned and ridden by Guy Weaving of Southmoor. Having already won the first race, he jumped the last fence in front and was beaten on the run-in by Darby Butt in the second, over the Old Berks course on land farmed by him at Middleleaze, Coleshill.

I hope recitation of some of my family ancestry has not been too boring but I have tried, and probably failed dismally to describe the kind of era and background into which I emerged on 18 May 1913. Purposely I have only made scant reference to my dear mother who died in 1944, aged 63 and father who died tragically in a tractor accident in 1961 aged 73. I feel their immense contribution to my life and my feelings for them will emerge as I write on. O' that I was more appreciative in my earlier years. This failing of mine was, I think, first brought home to me, after I was married, when I chanced to meet a friend of the family, Mr. W. N. Chambers, a Faringdon Auctioneer, a few days after Pamela, our first child was born in 1944. The conversation went thus: "Ronald they tell me you have become a proud father" — "Yes, Mr Chambers" — "Well done, congratulations to your wife; you will now begin to realise exactly what YOUR parents did for YOU." I can only admit time has proved how right he was.

Father had an older brother as already mentioned, a younger brother Leslie, usually called Peter and two sisters Ethel and Edie. While my Mother was the middle one of three sisters, the other two being Edie and Eva.

Father and Mother were married at Marylebone Parish Church, London W. on 26 October 1911, by the Rev. R. Conyers-Morrell (I have his card!). The reception was held at The Grafton Hotel, described as being "opposite Maples" in Tottenham Court Road. The brochure card stated "It is most comfortably furnished by Messrs Maples and comprises 180 Bed And Sitting Rooms, Electric Light, passenger Elevator to all floors and handsome Public Rooms, Vacuum Cleaner".

Tariff of Charges — Bed, Bath & Breakfast (Attendance & Lights) from 5 shillings (25p in today's money!).

Table d'Hote Lunch	1/6 & 2/-	(7 ½ p & 10p)
Table d'Hote Dinner	2/6	(12½p)
Afternoon Tea	6d & 9d	(2½p & 3.3/4p)

The honeymoon was spent at Brighton, staying in the "Haslemere" Private Hotel in Montpelier Road; with similar facilities to the Grafton in London. En Pension terms were double room — 3 to 4½ Guineas per week or 6/6 to 10/6 per day inclusive of attendance and electric light. Bed & Breakfast 4/6 to 5/- per day. Lunch 2/-, Dinner 3/-. Such prices seem unbelievable today. I was at a race meeting a few months ago, felt like a cup of tea at around 4 o'clockish. The lady in the tent close to the paddock who served me said "That will be 75p luv," and I couldn't resist saying "if my mother knew I had paid you 15 shillings for a cup of tea which used to cost her tuppence she would come and haunt you for the rest of your life". "I know luv ain it awful," she replied in a tone that indicated she felt the same as I did, but couldn't do much about it!

With their honeymoon over I presume they took up residence immediately at Coxwell Villa, where 19 months later I was to be born. I also presume father carried on business as usual in partnership with his brother Stanley who had also married a Miss Porter from Swindon and moved in over the shop. It must have been around this time that the firm acquired its first motor car. I was told it was a Swift with a folding canvas hood and what were then called bucket seats. No doubt this was a great asset, as father spent a lot of time on the road, calling on customers to take orders, a task that had previously been accomplished by pony and trap. Most grocers in those days, indeed up to the early 50's, wholesalers and retailers alike, employed travellers to call regularly on the larger houses and farms in the district, the goods ordered being delivered on the doorstep, upon a particularly appointed day in the week, accounts being settled monthly. How different today with the existence of the vast numbers of "If you want it, come and get it yourself" stores. I often wonder whether it really is such a saving to the consumer as it is supposed to be, when time, cost of petrol and perhaps a certain amount of unnecessary purchasing is taken into account. Judging by the sight of some of the car parks provided there might not be so many vehicles on the road either. It's only a passing thought!

TWO
World War I: 1914-18

Back at Coxwell Villa any recollections of babyhood are practically nil, although dimly in my mind I can see a few men dressed in loose fitting pale blue suits with cream reveres to their jackets and wearing narrow red ties which I came to know later as some of the poor chaps who had been wounded in the early part of the First World War and were being cared for at Kitemore House which like other large houses had been taken over by an army nursing unit ably assisted by local ladies of the V.A.D. of whom father's sister Edie was one. I gather Auntie used to organise small tea-parties for those who were able to get around with or without crutches.

It must have been when father was called up in 1915 that we left Coxwell Villa and moved to "The Ferns" in London Street, close to the middle of Faringdon. I should imagine the move was made so that mother could be in closer touch with her friends, while father was away. Due to medical rating and partial blindness, he never served abroad. Despite requesting posting to the Royal Army Veterinary Corps bearing in mind his love of horses he finished up in the R.A.M.C. the Royal Army Medical Corps, known by most soldiers as either the 'Rob All MY Comrades' or the 'Run Away Mother's Coming' oufit! Most of his service was spent at Portsmouth and Blackpool. In his own words his job was "to help get other poor devils fit to go overseas".

Father did, of course, come home on leave but occasionally mother would take me to Blackpool for a fortnight or so when and where she was able to find suitable accommodation in boarding houses etc. It was then, I suppose, at about the age of 3½ years that various incidents and happenings began to store themselves away in my tiny mind, as they do to almost everybody else and it really is amazing how we can think back seventy, eighty years or more and recall some of those incidents in detail, while quite often it is difficult when one grows older to recall what happened a few days ago. I suppose we were much more impressionable when young or maybe the storage space in the old brain-box begins to run out of pigeon-holes!

Although I have never been to Blackpool since, there are one or two recollections I have. Firstly, turning out of a side street on the sea front, being almost picked up by a gale force wind and seeming unable to get my breath back, until father pulled me into the shelter of his army great-coat. I probably thought I was going to die! I remember having a ride on the Big Wheel; my impression of it today being that it was a much larger version of the present day fair ground wheel, with seats for two. The Big Wheel took you round in what was like a Wendy House, I think there were a dozen or more carrying about half a dozen people in each.

The Blackpool Tower was probably the highest thing I had seen up to then. I don't recall being taken up it, but I do remember being put to bed one afternoon to rest, something mother did, if they wanted to go out in the evening. There were no baby-sitters then, so they had to take me with them! The object of the exercise on this occasion was a visit to the famous Tower Ballroom, made even more famous latterly by Reginald Dixon who thrilled radio listeners with his playing of its mighty Wurlitzer organ. The picture in my mind is that the interior was something like the Albert Hall, the huge floor space in the centre being covered with dancing couples and most of the men wearing Army or Navy uniforms. Occupying a small area in the middle of the floor was a square carpet on each corner of which stood a uniformed military policeman with cane under arm, for the purpose of making sure the rules of reasonable behaviour were being obeyed. It would be interesting to go back again after the passage of some 70 years just to see if there is any resemblance to the scene I have described, but don't suppose I ever shall.

I was also taken by mother on one or two visits to London where we stayed with Granny Greenip, in the Kilburn area. It was on one of these occasions, that the whole household which included the "inhouse" staff of the Bon Marche department store for whom she was housekeeper, were awakened during the night by the night watchman and in our dressing gowns, were shepherded below ground into a cellar, as calls of "take cover" had been heard in the streets outside, indicating that German Zeplins were overhead and that bombs could be expected to fall anywhere. In fact, apart from a distant rumbling of anti-aircraft guns nothing else happened and after 2 or 3 hours, a lot of chatter and quite a few cups of tea, the streets echoed to the sounds of "all clear" which I have a notion came as did the warning to "take cover" from bugle calls from senior boy scouts riding the streets on bicycles.

I think history records that Jerry's Zep raids were not very successful. I believe one was brought down in flames and although they did manage to drop a few bombs on the capital and sadly a few people lost their lives, the threat never developed into anything likely to effect the outcome of the war. Perhaps they thought it was a useful attempt to put the wind up people; if they did, then they greatly under-estimated cockney courage, as they did again some thirty years later.

Other incidents also come to mind during the six years spent at Faringdon before moving to Baulking. One was being taken to see an aeroplane for the first time. Due to some reason or other it had made a forced landing close to where the Old Berks Kennels now stand, at the bottom of Jaspers Hill. Having come to rest without too much apparent damage half way across a hedge and ditch, it was there for quite a while and attracted a large number of sightseers. I don't remember too many details except that it was a bi-plane, silver-grey in colour with wings that were spaced apart by wooden struts and secured by crisscross wire cables. The wings and body fuselage carried the red, white and blue roundels of the Royal Flying Corps as it was known before being renamed the R.A.F. People were quite excited about it all, especially as the pilot seemed to become quite a hero. The sight of an aeroplane in the sky in those days was rare and quite something, the noise of the engine bringing people out of their houses to gaze and marvel.

I am reminded of a relevant story told to me by the late Jack Roberts who lived at Ashbury and worked on the maintenance staff of the Craven Estate. He fascinated me often with true stories which he told in real Berkshire dialect, like the old shepherd who described seeing his very first aeroplane and how it crashed near to where he was tending his sheep on the downs somewhere above Ashbury. With a sweeping wave of the hand to indicate flight. "Over er cum, all of a glitter," he said "the sheep run up the vold an knocked t'hurdles over, the 'oss rurred upright in the sharves o' the warter-barrel, an all of a sudden, wump er went a Frogley's Linch". Linch is apparently an old name for ledge so I imagine the plane crashed on a nearby ledge-shaped piece of downland which had something to do at sometime with a Mr Frogley! Maybe it is recorded somewhere on an O.S. map for the area.

Jack Roberts also told of another old shepherd who, returning home from Ashdown at the end of the day, came across a lone walker resting on the roadside verge to admire the marvellous sight of the sun setting over the Vale from the top of Ashbury Hill. Following a few friendly exchanges, the walker inquired if the hill was dangerous. "No Sir, not up yer it ent, tis at the bottom, wur they all seems t' get into trouble" was the ol' man's reply.

Jack mentioned at one time there were eight full time shepherds living in Ashbury. Today I very much doubt if there is one and indeed a folded flock of sheep is a very rare sight anywhere these days. He also said most of these characters met regularly at The Rose & Crown for a little liquid sustenance and a game of dominoes. On more than one occasion he remembers one particular member of the circle getting up from the table to refill his mug and knocking the table leg with his foot sufficiently hard enough to disarrange the pieces already played, having decided without doubt that he was going to lose. Accusations that he had done it on purpose were always met with "T'was an accident, les not argue, fayers fayer les start agin"!

Reverting to Faringdon, something happened while father was away in the army, that could have meant early curtains for me.

I suppose I was about three or four, still in the push chair age anyway. Great Grampy George had told mother there were a lot of raspberries that needed picking in his walled-in garden tended by his gardener in Southampton Street, which could have been where the fly-over near the car park is today. Mother was told to help herself, so with jam making in mind, I was taken along in the push chair and was having great fun on my own, running round the garden paths encompassed by small trimmed box hedges. This was fine for a while but later Mother's shout from the raspberry canes, as to where I was received no reply. Panic stricken she apparently tore round the garden shouting "Where are you"? — until in the middle of a path she came across an old wooden cover off a well which I had managed to push aside and had fallen into. It was fortunately only four or five feet in depth I was told later but deep enough to have drowned me if it had been anywhere near half full. That it wasn't was due to the fact that it was a dry summer. The well was used for the purpose of collecting surface rain water to which was added the odd barrow-load of manure to produce a beneficial liquid with which to encourage the various crops being grown nearby! Apparently, this was normal practice in those days and lots of old gardens had similar facilities. I dare say the modern procedure is to mix it all up in a tank above ground. I'm no gardener, so I wouldn't know, but its got to be safer! Anyway mother described the whole drama to me in detail more than once. How she lay face downwards on the path in her light summer dress, managed to fish me out and not smelling particularly like a bunch of violets, rushed me home in the push chair to get me into a bath as soon as possible.

Another incident which comes to mind concerns a trip to Wantage for an appointment with Tom Reveley, the photographer, to take a family photo *(reproduced herein)*.

By the look of the result we (Mother, Father and I) were attired in our Sunday best. The journey was made in a recently acquired "Hupmobile" motor car — whether it replaced the "Swift" or it was an additional vehicle I know not, I do know however, it was primarily to do with the grocery business because one of the doors displayed the name Liddiard Bros., Grocers, Faringdon, over which a motoring rug was "carelessly" draped on journeys other than those for which, I presume, it was licensed! Be that as it may, about half a mile out of Faringdon at the top of Folly Hill a puncture occurred in the off-side rear tyre. Now instead of changing the wheel completely as today, the car was jacked up sufficiently to enable the outer metal rim to which the tyre was fixed to be removed by the unscrewing of a number of nuts. The spare rim with tyre already affixed was then pushed into place over the wheel centre, the nuts tightened up and hey presto, off you went. However, it was possible that father hadn't quite got it right, because about two miles further on, along the straight piece of road near the present day rubbish tip, we were overtaken by the spare tyre and rim, which finished up by careering into the wall on the right hand side of the road, some hundred yards ahead of us, much to my amusement but nobody else's! — as it was deemed necessary to postpone the mission and try again another day. By the way this type of spare wheel was called a Stepney. I don't know why — I think I heard father call it something else, but at that age of course there were some words I didn't understand.

During the last two years at Faringdon I attended a small school run by a Mrs Bye at her house in Gloucester Street almost opposite the entrance to the present County Library. I can't remember much about it — perhaps I wasn't particularly interested but I do recall some of the excitement when she explained to us that the war with Germany had ended, that peace had been declared and the soldiers would soon be coming home. At five years old, obviously I was unable to understand the reality of what was happening. It seems unbelievable now that by the time I was thirty it was all going to happen again, just because some little German painter and house decorator had the bizarre notion he could subdue and rule the rest of the world.

The only child I can recall being at school with me then, was a little girl named Hughes. I can't remember her Christian name, but she was a daughter of a builder and plumber who lived three or four doors away from us in London Street. She was my first real playmate; quite a stocky little lass she was and we spent quite a lot of time in one-another's houses. I remember they had a five star swing in their garden; while mine was only about three! As I was an only child I suppose mother considered a bit of feminine company was good for me! I'm sure she was right!

It was in Mr Hughes' office that I first saw a telephone. It was a wooden box affair fixed to the wall with a mouth piece sticking out of the middle for speaking into and a separate ear piece hanging on one side. The service was set in motion by turning a small handle on the opposite side of the box, this alerted the operator at the Post Office in Marlborough Street who did his best to get a line to the number requested, which sometimes took quite a while. I did in fact have the ear-piece to hold to my ear for a moment while Mr Hughes mystified me by saying with a grin on his face that there was a little man inside the box that did the talking.

There were quite a few Hughes in Faringdon at that time, the milkman, known as "Skimmy"! who used to drive round the town in his purpose built yellow milk cart shouting "milko" to bring householders to their doors, jug in hand, into which

Family Portrait

*Baby Ron Liddiard
dressed in traditional frock of the period,
with his parents, Kitty and Cyril Liddiard.*

was tipped the required measure dipped from a large lidded bucket with long handled standard ½ pint, 1 pint or quart measure. The product was liable to be sampled at any time by the Police and story has it there was a milkman in Swindon who regularly fell over in the street, bucket and all at the sight of one.

Then there was George Hughes, a kind of handyman who did all the local bill posting, riding round the district on his bicycle, with the paste bucket hitched on the handle bars — and a long handled brush tied to the cross-bar of the bike. Ably assisted by a nice ol' boy, Minnie Goodman and a Verney Gerring, he was employed by Hobbs and Chambers to erect all the hurdles etc. on Market days and clear up afterwards. I also recall George Hughes doing a bit of "town crying" but not in uniform.

His wife, a kindly soul, kept a small sweet shop in Gloucester Street, opposite what was for many years Langfords, the Corn Merchants, having been Fred Porter's before that. The latter name can still be deciphered high up on the building. I did in fact spend some of my very first pocket money with Mrs Hughes, being very partial to what I think were called "Sherbet Dabs" which consisted of a small yellow paper bag containing sherbet and a short stick of liquorice which was dipped into the bag to enable the sherbet to be sucked off it. They were only a half-penny each and I can almost taste the mixture of liquorice and sherbet now!

With the war over, it was back to civvy street for father and his two brothers. Father's army medical rating of C3 indicated, as well as being almost blind in his left eye, due to a childhood accident with a pair of scissors which left a scar on the pupil, he was also suspected of having what he called a "dodgy" lung. Although despite being a fairly heavy cigarette smoker it never seemed to affect him in any way. He was a small, tough resilient man and I recall him being exceptionally free from illness. If he had a cold or a touch of 'flu he just seemed to smoke a few more cigarettes!!

Anyway, Grampy was well aware of the current medical position and also the fact that most of the time my Father and his brother Stanley were in business together, they got on like fighting-cocks. So Stanley being the eldest took over the grocery business and father was installed as tenant of Church Farm, Baulking, a property which became vacant upon the death of George Collins in 1918 and Grampy purchased at auction for little more than £3000.00.

Together with about 50 acres of adjoining land which he already owned, known as Forty's, where I now live, this made a useful parcel of approximately 180 acres, 160 of which was very old permanent pasture. According to father's first year account book the annual rent payable to Grampy was £275.

To complete the post war re-organisation, father's younger brother Peter, who was a bachelor, still living at home, became Grampy's right hand man at Steeds Farm, Great Coxwell.

Pete, in fact saw more of the war than his brothers, serving with distinction in France, and being awarded the D.S.M. I remember Granny displaying the medal and ribbon with pride before it was framed and hung in the sitting room at Romney House.

THREE
Move to Baulking

I wasn't old enough to know how mother felt about moving from town to a country hamlet, composed of four or five farmsteads and a few cottages bordering an 18 acres village green, without a shop or pub but, I do remember her saying a few times later, perhaps not too seriously, that she married a grocer not a farmer! I'm not sure how father viewed the change either but I suspect he probably relished the prospects, as he had moved around in farming circles, sampled some of the graft involved at Steeds Farm, was mad on horses and hunting, liked a bit of dealing and before the war had already ridden a point-to-point winner or two, notably, a horse named Laurel which belonged to his sister Ethel's father-in-law John Wheeler who farmed at Common Farm, Uffington. A newspaper cutting dated May 31, 1911 describes the event admirably and I have a photograph of horse and rider taken outside Common Farm (again by Tom Revely) a few days after.

I recall father telling me there were great celebrations in Uffington during the evening of the win, which took the form of plenty of free beer and the appearance of the now defunct Kingston Lisle Brass Band marching round the village.

The Band was composed of about a dozen locals led by Bandmaster Dicky Eldridge. They performed at functions including visits to nearby villages at Christmas for the purpose of playing carols outside the homes of local folk.

The headquarters of the band was a room at the rear of the White Horse public house at Uffington, where the instruments were stored. Band practice took place on Friday evenings followed no doubt by a few pints of Morlands. One particular evening, Dicky Eldridge was prevented from arriving in time to open the proceedings so those already present started as best they could, only to be interrupted by Dicky opening the door to say, "Sorry, I'm late, chaps. I've never heard such a row as you are making; it would do you all good to come out here and listen."

The stage was set for us to move into Church Farm, Baulking on Lady Day, 25th March 1919. An early inspection of the farmhouse had revealed a great deal needed to be done by way of alterations and decoration. These were put in the hands of a Mr Wheeler who owned and lived at The Brickyard close by Uffington Station. He also had a sizeable builders business at Wantage from which he master-minded the building of what became known as the White City on the western outskirts of Wantage to where they had retired on leaving The Brickyard — some of which later.

Equipping Church Farm with live and deadstock was obviously quite a task for someone who had been used to stocking up a grocery business. It is interesting to note from father's account book that the in going valuation in April 1919 for livestock was £223 and deadstock £130. I have picked out a few details and prices of some of the additional equipment and livestock required to get the enterprise up and going as quickly as possible.

They were as follows:—

Implements — Horse Drawn		£. s. d.
Single furrow plough	New	8.10. 6.
6 R.N.F. plough shares	New	10. 6.
Set Chain Harrows	New	8. 0. 0.
Set Zig Zag Harrows	S/H	3. 0. 0.
Set Heavy Horse Harness	S/H	2 15. 0.
Knapp Horse Hoe	New	6. 0. 0.
Horse Rake	S/H	11. 0. 0.
Horse Rake	New	16. 0. 0.

Mowing Machine	New	24. 0. 0.
Massey Harris Hay Loader	New	38. 0. 0.
Martin Side Rake	New	25. 0. 0.
Bunce Cultivator	New	16.12.6.
Roberts geared elevator	S/H	10. 0. 0.

Barn Machinery:

Amanco Petrol Engine	New	30. 0. 0.
Bamford Root Pulper	New	28. 0. 0.
Cake Crusher	S/H	6. 0. 0.
Chaff Cutter	New	25.11.6
Milk Cooler	New	9.12. 6.

Livestock:

Young Bull	£47 (Bought of his Uncle Andrew Chillingworth)
Cows with calves	Av. £35/40
Heifers	Av. £30/35
1 Sow & 9 pigs	£35
2 Hens & chicks	£2. 8. 0.
A Gander	£1. 5. 0.
4 Ferrets	11. 6.

Feed-Stuffs according to sample quality were:—

Wheat	£20 per ton
Barley	£18 " "
Oats	£19 " "
Bran	£16 " "
Toppings	£15 " "
Cattle Cake between	£21. 5. 0. & £24. 2. 6.
Wheat Straw (loose) around	£2. 12. 6 per ton
Barley Straw (loose) "	£1. 12. 0 " "

Other figures illustrate a little of the financial side of father's first three accounting years, namely:

	Expenditure	Receipts	Labour Costs	Milk Sales
Y/E 31 March 1920	£5064	£2994	£483	£1395
Y/E 31 March 1921	£2836	£2507	£655	£1470
Y/E 31 March 1922	£2177	£1843	£475	£1187

As to the labour costs, 5 men were employed, a cowman, carter, two general workers and a groom/handyman so the average wage for the years 1920 and 1922 works out at roughly thirty-eight shillings a week for each of them, which included overtime, the only statutory holidays then being Christmas Day and Good Friday! Maybe in the year 1921 father took on an extra man, which could explain the increase in labour costs for that particular year, but I don't know. I do know though that the equivalent rate today would be £1.90p per week — little more than the present cost of a pint of lager!

With regard to milk sales which were the main contribution to receipts, there is no record of gallonage sold or the price per gallon received but I'm fairly sure the herd size was around 30 — 35 cows, assuming an annual lactation of around 750 gallons per cow a quick calculation reveals an average price for the year of a little over a shilling per gallon — 5p in today's money. Which has conjured up another statistical fact in my tiny mind indicating that in 1920 it required the production of 38 gallons of milk to pay the cowman's weekly wage while today it requires at least 384 gallons to do the same.

I suppose an economist would explain why this is so but it does seem it might have something to do with the high prices of which some products have soared. Its just a thought!

Words of that great farming character the late Tom Parker, who farmed in a large way at Fareham in Hampshire are beginning to ring in my ears. He once said to me when we were on an "Any Questions" panel and I had supported an observation with some figures, "Don't forget my boy, all figures can lie and all liars can figure"! Having mentioned his name I must tell a story or two concerning him before I move on. Like most farmers Tom very much enjoyed looking round other peoples' farms. On the occasion of a day spent for this purpose with another well known farmer, the late Dick Roadnight, at Britwell near Wallingford they arrived at the back door of Dick's farmhouse. Getting out of the Land Rover, Tom asked for the broom to brush his boots prior to going indoors. Dick answered

"We don't use a broom for that Tom, I thought you were up to date down Hampshire way, there is a rotary brush in the porch — all you have to do is turn the handle and stick your foot out"! One up to Dick one might have thought — but was it? Some months later when Dick's hospitality was being returned, Tom had installed a similar piece of equipment but operated by a small electric motor and at the appropriate moment said "The brush is in the porch Dick just press the button" — All square.

The other story concerned a carter employed by Tom during the cart horse days. While riding round on his cob one morning to see how the carters' ploughing was progressing Tom stopped on the headland to have a word or two. On leaving, the carter asked if he could have a couple of hours off next Friday afternoon. "What for?" asked Tom; "To get married" was the reply. "Well," said Tom, "You better have the day off hadn't you?" "No Sir, thank you, a couple of hours 'ull do if thee can get the shepherd to lean on this yer plough while I be gone, I can be back to hitch off and feed round afore night" was the carter's answer. Tom told me the marriage foundered a bit after a few months with the result that his wife left for a while, only to return to him later. Having heard the good news, Tom said he told the ol' boy how pleased he was to hear his wife was back with him, which brought forth the reply, "O arr 'ers back alright, an I give er a good hiding too". "What, for going?" inquired Tom. "No, for coming back," said the carter!

I also remember Tom proposing a vote of thanks to Lord Huntingdon who I think was under-secretary to the Minister of Agriculture, Tom Williams during the last war. The occasion was a large gathering of farmers invited to Winchester, in about 1941 for the purpose of listening to the Government's exhortations and proposed methods of encouragement to persuade farmers to use the plough for the purpose of producing as much as possible from the land within our shores, especially wheat, which prompted Tom, after pledging farmers whole hearted support, to finish his expression of thanks to Lord

Huntingdon for addressing us with the impish suggestion that, "If the Government offered to pay us a bit more then I'm sure his Lordship would have more wheat than he wants"! There is a little more I could say about Tom Parker, his love of horses, his coaching expeditions around the English countryside staying with farming friends overnight, competing from the driving seat in "Four-in-Hand" classes at numerous shows up and down the country, his strong team of matching piebald ponies that once pulled the milk floats around the Portsmouth area selling the milk from his dairy herds. I remember him showing them to my father with great pride when he hosted a day's visit from a large number of members from the Faringdon Farming Club. For those who like reading about countryside characters like Tom Parker, a book has been written about him, entitled — "Tom, Tom the Farmer's Son" — Its very good too, but I had better get back to Baulking.

The land at Church Farm is typical of that lying within the Vale of the Uffington White Horse. Being fairly flat, the soil is quite heavy with a ploughing depth of about 8 or 9 inches and overlies a great depth of almost impervious clay sub-soil, so it doesn't drain very well. The land when father took over was contained within eleven fields, 18 acres of arable, which was known as Wheat Close, and the other ten rejoiced in such names as Maids Mead, Goose Green, Anty' Leas, Banny Hill and the like, the biggest being Home Ground which was 42 acres and had an unfenced gated road running through the middle of it. In those days there was another gate at the southern entrance to the Village Green, with a kissing gate to accommodate those using the foot path running along the side of the road. There being very few motor cars around, the road was used mostly by horse traffic and if travelling towards Stanford in the Vale the three gates all within the space of a mile were a bit of a b.... especially if you were travelling on your own. It could be difficult to keep the gate open while getting horse and carriage through, then having to turn round and hold the horse while struggling to shut the gate to keep the cattle in. During the winter, when cattle were yarded, the gates were either taken off or tied back.

Quite often during the summer, if the traveller was lucky he would find a few children playing close to the three problem areas and there would be a mad rush by them to open the gate in the almost certain hope that they would be rewarded with a penny and if they were not then the traveller was sort of black listed and any following appearances by him ignored. I know this to be so as I had the pleasure of a share of the takings more than once!

The farming method was quite simple. The small arable acreage adhered more or less to a four course rotation of wheat, oats undersown with red clover which usually gave two crops of hay, was then mucked with farmyard manure, which was carted out into long lines of small heaps about 8 yards apart from which it was spread by hand forks. It was often called "dung spurling". This having been done it was ploughed in for cropping with roots, usually a split acreage of mangels and swedes and perhaps a few turnips. This root break served as a cleaning crop for the two cereal crops which followed. It must be remembered that there were no sprays of any kind in those days.

During the summer the dairy cows spent most of their time on the 42 acre field closest to the farmstead without a change of grazing until the aftermath had grown on the rest of the fields which, bar one or two small fields accommodating the young stock, were mown for hay.

FOUR
School Days

I was nearly seven years old when starting my first term at the village school, which was only a couple of hundred yards away on the edge of the Village Green close to the main London to Bristol Great Western railway line. The headmistress was a Miss Fricker and assisted by Miss Vera Wheeler, a daughter of the builder, they taught about 35 children of all ages, up to 14 I think it was, who lived within the parish boundary, the farthest away having to walk 2 miles from Oldfield Farm cottages in all weathers to get there. Not many folk, certainly not children, owned a bicycle then and I can't remember anyone arriving on one other than Vera Wheeler the junior class teacher. So mostly, it was "Shanks' Pony" for work, school, Church and pub, the latter being about a mile away and of course a mile home. Known as "The Junction" it was at Uffington Station owned by Halls Brewery of Oxford. The landlord was Bert Tame, a nice chap who enjoyed playing football for Uffington and was quite handy as a carpenter. I remember him doing a few jobs for father including making wire netting guards for the south facing windows of the farmhouse when the front garden was turned into a tennis court. The lower ones were portable but the upstairs remained fixed until the summer was over. Bert Tame moved to Faringdon later to become mine-host at The Swan.

Almost all the parents of children attending school worked on the eight farms in the parish which all had dairy herds; now there is only one which is at Baulking Grange. There were exceptions, the Essex and Thomas families whose fathers worked as porter and signalman respectively for the G.W.R. at Uffington Station.

Organised games were not in the school curriculum so consisted mostly of impromptu football on the village green with caps or folded jackets put on the ground for goal posts. There was always an argument as to whether the ball was inside or outside the jacket or it hit the post, so to speak. Without a crossbar of course, there were often shouts of "too high" followed by similar argument, but it was all good fun. During the summer it was "rounders" and one or two attempts to play cricket on selected parts of the village green that had not been poached by cattle or horses and were relatively free from cow clats. Rather more serious attempts to play the game were made at the "bottom end" of the green — we were known as the "top end". I think the descriptions originated from living above or below the church. Down there, the so called "square" occupied the same area every year and was in front of Manor Farm, it was usually chain harrowed and flat rolled in the spring to level it off a bit and geese belonging to the Reade and Matthews families helped to keep the grass reasonably short. Sometimes the old ganders could be a bit tiresome and had to be rounded up. As no pads or gloves were worn it could be a bit "hairy" too! This was a game for the men, but youngsters were encouraged to join in and when they did so. The bowlers were told by those fielding to "go steady". Rather like the old veteran who turned out at Shrivenham after the first World War, in a match to help get their Cricket Club going again. One of his old mates was umpiring and when he got to the wicket to bat, being anxious to give him guard, shouted "What do you want Charlie?" "Better 'av a slow 'un to start with Bert," was the reply.

The village school, sadly, has gone the way of many other in our countryside. I say sadly because I feel schools and churches have been the centres of village life and communities for so long, that if the move to centralisation continues, the breaking up of local populations will mean the eventual loss of community spirit altogether. The same thing is surely happening with the disappearance of village shops where a smile and few minutes gossip can do a power of good to many folk. Much more I guess than driving perhaps ten or fifteen miles to park in a vast carpark, wander through a maze of aisles displaying thousands of everything, passing faceless people and finally queuing to see your money taken by someone who has hardly time to say thank you; to be eventually swallowed up by an organisation that is making millions. They do say big isn't always best. I suppose though we humans are a bit like cattle which adapt to the environment in which they are housed and managed; so perhaps the change-over will prove to be painless and go un-noticed, but I'm not sure about the consequences. — End of Party Political Broadcast.

The demise of Baulking School in 1931 was basically because the village had become somewhat of a bachelor's paradise; which meant that a period of time had developed during which an insufficient number of children were born in the parish to warrant the school remaining open. When I attended some ten years previously there were more than thirty pupils on the roll-call, but they all grew up and there was no one to take their place. It was therefore decided that the £216 of ratepayers' money being spent annually was too high a price to pay for the education of seven pupils, only three of whom lived within the parish boundary.

The decision by the Berkshire County Council Education Authority to close the school and direct the remaining pupils to Uffington, two miles away, was announced but not without causing a certain amount of animosity. A petition was also drawn up by the parishioners requesting that the school should remain open, but met with no success.

The school house and its two classrooms were then used to house a succession of School Attendance Officers, until the County Education Authority finally decided to sell it on the open market as a private dwelling.

I recall the years between 1919 and May 1922 as being a

happy time all round, the war being over, having brought obvious relief to the grown-ups, the challenge of father and mothers' start in the new business of farming and my first real taste of actually living on a farm with the animals around was exciting for me although I must admit I didn't take to it like a duck takes to water. It took time and I remember feeling very insecure with the larger stock such as cows and horses when in close proximity and made sure I had the company of some of my village playmates. Although they were of similar age to me I knew instinctively they were a damned sight more fearless than I was. Eventually I suppose this rubbed off on me and gave me the confidence needed to help transform me from a "townie" into something of a countryman.

Another thing that bothered me at the time was a kind of fear of being alone; I seldom was of course but I suppose the feeling came from being an only child used to living in a small town house with street lighting outside and people being around most of the time and the change to living in a much larger rambling old farmhouse, lit during the hours of darkness by oil lamps and candles and hardly being able to see your hand in front of you outside unless the phase of the moon and an almost cloudless night sky coincided and allowed the "parish lantern" as the moon was called, to shed light on the village and surrounding countryside; and what better sight from a bedroom window on a frosty rimy night when the moon is full and high and the sound of an old barn owl hooting in the distance can be heard.

Up to about the Second World War almost every farmhouse employed various forms of domestic help, such as housekeepers, maids, nurse maids, washer women etc. Due to prevailing circumstances there were very few opportunities for girls leaving village schools to find employment other than in the roles I have mentioned. The first to help mother, which also entailed keeping an eye on me — was Edie Ballard. She was one of a family of six, all of whom had attended Baulking School and whose father worked for the Stevenson Bros at

Oldfield Farm. She was, like the rest of her family, very small in stature with a kind and willing heart. I would think she was about eighteen when first coming to live-in at Church Farm and her weekly remuneration would have been about five shillings per week, with mother providing keep and a couple of sets of white cap and apron. Edie was like a little mother to me, having been the eldest of her family she had no doubt gained first hand experience looking after her five younger brothers and sisters so it came naturally to her. The only times she ever got really cross was when my cousins from Croyden, Jack and Bob English, who were my age, came to stay during the summer holidays and the three of us ran riot a bit, especially at bedtimes, if she had been left to see us to bed while our parents were out for the evening. Reminiscing about this many years later she said "I wish I had as many pounds as times I had smacked your bare bottoms, I should have been quite rich".

Edie fell in love with one of fathers' employees named Tilling, who lived at Uffington; he was always known as Copper Tilling because of the colour of his hair. I liked Copper too and used to spend some of his dinner hour watching him make a small cricket bat which he offered to shape for me by the use of a spoke-shave out of a piece of seasoned withy rail. The same sort of wood that home-made farm gates were made of during the winter when the weather prevented work outside. The bat was a great success and lasted quite a while before splitting.

Edie stayed with mother until marrying Copper and moving into one of the cottages attached to the farm, where their two children were born before they moved to Stanford-in-the-Vale to work for someone else. Sadly Copper died at an early age but Edie lived to be over ninety and died only a year or two ago. I am pleased to say I still see her son Bob occasionally.

During the rest of the years preceding the Second World War I can recall at least half a dozen lasses who helped mother in this particular way. In fact one was a younger sister of Edie's named Ethel who almost gave mother a heart attack one afternoon. Apparently they were spring cleaning the larder when there was a knock on the front door. Having answered the knock Ethel returned to the larder to say "If you please m'am its the Duke of York." —"What do you mean, Ethel — the Duke of York?" uttered mother in amazement, almost dropping a bowl of eggs. "Yes," said Ethel. "He's called to see if we've got any rabbit skins". There was a Mr Ginandria who kept the Duke of York pub in Faringdon (it stood on the present site of the United Church) and he diversified a bit, by combining an ice-cream round with the business of collecting a few rabbit skins at the same time. Ethel and the villagers all knew him as the 'Duke of York'. Incidentally, the skins were usually hung out to dry somewhere outside and were worth a penny or halfpenny according to the quality. A whole rabbit was worth between sixpence and a shilling. They were quite tasty with a bit of pie crust round them or in a stew — but I have never fancied them since the 'mixie' outbreak some twenty years ago.

The Duke's ice-cream business was conducted from the middle of a low, small wheeled, gaily decorated purpose built ice-cream cart with a fringed roof on top and penny cornets or tuppenny wafers could be purchased from the sides or rear. The outfit was drawn by a quiet old pony with a few bells attached to the harness to attract the necessary attention when trotting through a village and could be hired by arrangement to attend local fetes, flower shows and the like.

There has, over the years of course, been a succession of many employees at Church Farm, it would be interesting to know just how many. It is possible I could say something about most of them and no doubt they all left their mark in one way or another, but I shall have to content myself by picking out one or two as the years progress.

One of father's earlier carters was Bill Mace who came to work for father soon after he had been de-mobbed, having been with the Royal Artillery in France acting as a driver in a gun-carriage team. He was a great character and a loyal worker, even if his fuse seemed to be a bit short at times. With his wife

Nell and their three sons Bill, Cyril and Reg, all village playmates, they lived in a cottage close to the gate at the southern end of the green. Bill was its very last tenant, for the reason that it had been in need of repair for a long while, eventually becoming unsafe to live in. After the family had moved to one of Church Farm Cottages, it was partially demolished by the Village Green common holders for safety's sake. The small site and garden became overgrown and so it remains today and I'm not sure if anyone knows who owns it. It was a very tiny cottage and despite its state Nell always strived to keep it as spotless as possible, she was a great one with the scrubbing brush. She used to do a bit of it for mother too, Mother used to say "My word that woman has got some elbow grease, it wouldn't take her long to scrub a hole in a flag-stone".

Visits to the cottage when playing with the boys always fascinated me, especially if Bill happened to be sitting by the little kitchen grate, their only source of heating and cooking. On the walls were one or two army photographs of Bill riding the near side leading horse in a team of four, hitched to a gun carriage complete with its trailing cannon. Best of all, though, was the revolver hanging over the mantelpiece, which he said he took off a German officer, after knocking him down with the butt of his rifle toward the end of the war, then managed to smuggle it back to Blighty. What stories he told us. Sadly, his eldest son was killed in the Second World War. Reg, the youngest came through safely but died a while ago and Cyril died at Stanford recently. We often enjoyed recalling some of those bygone days. I shall always remember his father telling me more than once that he taught my father everything he knew about farming, after he came to Baulking. I think that was a bit of an exaggeration, but he certainly taught me, when I grew older, some of the crafts of ditching, hedgelaying, thatching, post and railing and many other tricks of the trade and often gave me a haircut for threepence which was half the going rate in Faringdon or Wantage.

The vicar of our small 13th Century Church of St Nicholas

at the time of our arrival in Baulking was the Rev. John Southby, a rather frail old gentleman and I don't think his eyesight was particularly good, as quite often his wife would leave her pew to put him back on course after he had lost his place whilst taking the service. Rather like the old story of the parson with a runny nose which sometimes caused the pages to stick together when The Book was closed following the lessons. A few Sundays later he got to the end of the page with the words "and Moses was sick," turning over two pages stuck together he continued at the top of the next "and the lot fell on Aaron". On one occasion the Rev. Southby was taking early morning communion. The frost having cracked the heating boiler, the church was like an ice-house, so much so, that dear old Rev Southby collapsed from the cold on the alter steps, whereupon he was revived by Mrs Reade of Manor Farm, who had the presence of mind to administer some of the wine before wrapping him up in a horse rug and taking him back to the vicarage.

It was not long after that he was succeeded by The Rev B.M.Hawes, a younger man with a teenage family. He was nicknamed "Bummy" by the school-children because of his initials. I don't know if he was aware of being so called but he was a nice chap with an obvious sense of humour as he passed this on to his churchwardens in the vestry after a morning service. He had been conducting a course of evening confirmation classes at the Vicarage. At the conclusion of the last one, he summarised his efforts to explain to about a dozen candidates the reasons for being confirmed and how it would strengthen and benefit them in their spiritual life. Being the end of March the Boat Race was very much in peoples' minds, so he used it to illustrate his point by saying that similarly the crews prepared themselves by training hard and eating a lot of beef-steaks etc. to fortify themselves so that they could row faster and win the race. Feeling that he had prepared them well and that there was little else he could say, he turned to one of them, who happened to be a girl and said, "Now why are you being confirmed? and she answered "To make us row better" I think there may have been a suspicion of mickey-taking in that remark.

While in the realms of Church stories there are two I would like to pass on, one true, the other is supposed to be.

The first concerned a great farming character the late Sid Reading. He farmed at Rectory Farm, Langford and was well known as a breeder of Lincoln Red Cattle. They were a dual purpose breed which declined in popularity following the second war when specific single purpose dairy and beef breeds became the accepted. He was a church-warden of his parish church and was asked by the vicar in the vestry immediately prior to morning service if he would read the two lessons as he had suddenly developed a very sore throat and it would be a great relief to him if he would? "Yes, of course he would." So, without knowing what the lessons were or any preparation, he was confronted early on in the first lesson with a whole host of biblical names which needed a bit of pronouncing; anyway he mumbled and coughed his way through them only to find them cropping up again later, so he quickly got round the problem by saying "and all those gentlemen I mentioned before".

The following is supposed to have happened in a local village school during a scripture lesson concerning the story of Moses being found in the bulrushes. Having explained it all and to make sure they had been paying attention, the teacher asked if anyone could tell her who was Moses' mother? Whereupon a bright lad from the back of the class stood up and said. "If you please Miss, it was Pharaoh's daughter." — "No, No, answered the teacher. "She was the one who found Moses in the bulrushes" — "Ah, so her said," the lad quickly replied.

The village water supply at that time consisted of a well and hand pump in the farmhouses which some thought provided water fit to drink while others didn't. The rest of the population made daily visits to the village pump, taking their requirements home in a couple of buckets suspended by small chains to a wooden yoke which fitted across one's shoulders. A small cross made of wood was floated on top of the water in each bucket to help prevent water splashing over the sides. Having been used to tap water in Faringdon, mother and father didn't take to the taste of Baulking well water, although those who had been drinking it for years seemed to have suffered no ill effects. However, there was a supply of spring water which years before had been piped by the Craven Estate from its source at Britchcombe which they owned to other parts of their estate in and around Uffington, one of which was Moor Mill Farm near Uffington Station, and a standpipe had been erected by the side of the road outside the farm for use of the cottages nearby. So after the milkcart had unloaded Church Farm's daily milk output contained in seventeen gallon churns on to the milk train for Paddington, an extra churn was "borrowed" when empties were doled out by the railway porter supervising the operation, and the milkcart proceeded to Moor Mill where the water was drawn from the standpipe into a bucket which was then tipped into the churn and taken back to Church Farm where it stood in the old cool dairy ready to be dipped into with a jug to provide us with enough drinking water for usually three days, after which the collection was repeated. It wasn't until about 1928 that the Faringdon Rural District Council negotiated a deal with the Craven Estate to allow them to pick up their supply in Uffington and bring a piped supply to Baulking on the condition that the water was only available from five standpipes, situated at vantage points on the village green and could be used only for domestic purposes. There was also a condition that the Faringdon Rural District Council should make the supply available to the Craven owned property Oldfield Farm but through a smaller pipe. The whole scheme was organised to provide relief for the unemployed in areas of South Wales, where the work was put out to tender. About half a dozen Welshmen lodged in and around the village while doing the job, which took a few months as the trenching was dug by hand tools and was more than three and a half miles in length. The completed service was an obvious step forward but for the "regulars" to the village pump the only advantage was that most of them didn't have to walk so far. In fact some still preferred the old source of supply and for sometime walked past the new standpipes to get it.

I am reminded of Grampy saying he once went to look over a small farm down Witney way for the purpose of a possible purchase and, when he asked the somewhat elderly owner what the water supply was like, he replied "I been yer a long time and d'know I can't 'onestly say I remembers tasting it neat".

After about three years at the village school it was decided that I should become a boarder at King Alfred's School, Wantage. I started there on 4 May 1922 and was nine years old a fortnight later. Arriving with Mother and Father in the early evening on the doorstep of Junior Lodge which was in Newbury Street, I remember bursting into floods of tears, whereupon Mother took me off for a walk up past the Old Comrades Club to the Recreation Ground to try and pacify and comfort me; bless her, I daresay she was feeling much the same as I was. I recall her telling me in later years that I said to her "If only you could stay with me for a night or two, I'm sure I should be alright and get used to it quicker". Anyway, eventually having met the housemaster Mr F.T.Brooks, who I soon discovered was known as "Brookie" and been introduced by him to the matron Mrs Parker — known as "Parkie" , Mother and Father left and I was put into the charge of one of the more senior of the twenty boys of the house, named Nunney, (initials A.E.) who took me under his wing, showed me the ropes and helped me mix in with the other boys a few of whom were from the Faringdon area. I shall always be grateful to Nunney for his kindness and by a strange coincidence a year ago, Michael Green, a current Joint Master of The "Old Berks" was telling me that the Hunt had received a silver cup with an accompanying letter saying that a relation of the donor had won it many years ago at an O.B.H.Puppy Show; on inquiring who the donor was, he said he came from Norfolk, and thought his name was Nunney which rang a bell with me, and I told him I was at School with a chap of that name and thought he later became Clerk to the City Council in Norwich. Michael 'phoned me later with the Norwich address and telephone number on the donor's letter and a further 'phone call by me revealed it was my schoolboy chaperon and that he remembered coming with me to Church Farm for lunch and tea on Exeat Sundays during term-time. Questioned as to how he came by the silver cup he told me his grandfather won it when he was mine-host of the Bear Hotel in Wantage Market Place just before the turn of the last century, when he hunted and walked puppies for the Old Berks, and Nunney felt as he himself was ageing and it was of little interest to others, he thought it would be nice for the O.B.H. to have it back again, so to speak. It is now awarded annually at the Puppy Show for 'the best working bitch of the season.'

K.A.S. as the school was often called, consisted then of 140 pupils, approximately 90 of whom were boarders and accommodated at the Priors Hold, a house close by the Parish Church, the housemaster being a Major Metcalfe; Major Allen was housemaster at Wessex House which was next door to Cleggs the Chemist and backed dangerously close to the perimeter wall of St Katherines', a Convent School for girls.

It was well known that the wall was not insurmountable; it was also well known by the inmates of Wessex that the feeling of Major Allen's cane across your backside might just prove to be a deterrent worth taking into account if a foray was being contemplated.

John Arbery was another old Alfredian who carried on a family drapery business in Wantage Market Place. It is the only shop I know which still has a fully operative compressed air pipe line connecting the shop counters with the main office for the purpose of dealing with sale notes, and purchase money. Having been recorded and dealt with, the receipt sale note plus any change due is returned immediately to the respective counter and handed to the customer. I don't know if I would be right in assuming that this was installed before cash registers came into regular use. Perhaps not.

Since recording this John has retired, put the shop and the business on the market and, to underline the great age and old fashioned character of the shop, Central Television thought it worthwhile to pay it a visit on the day it closed for the purpose of recording the closure on film which was included in their news bulletins for the day.

The rest of the boarders lived in the original main school which consisted of two main dormitories sleeping about 24 boys in each, the stairway to which bore the names of many former pupils which they themselves had scratched on the wood or written on the walls. With all the alterations that have taken place over the years I daresay they have disappeared. There was a dining room divided by an arch which seated about 50 boys and the school staff dined separately on the other side of the archway. All teaching was carried out within the confines of the school and boys living in houses returned there for meals and changing after games, on half days — Wednesday and Saturday afternoons. There was a fair sized Gymnasium devoid of any equipment, suitable only for drilling and floor exercises etc. The school assembly hall was used for religious services and was divisible into three areas if necessary for additional classrooms.

About the time I arrived a school chapel was built, the whole interior being of varnished match boarding, clad on the outside by green painted corrugated iron sheets and a felt roof. It was christened, a little irreverently perhaps, The Tin Tabernacle, and if it rained heavily during a service it was difficult to hear yourself singing, let alone praying. Nevertheless I recall vividly some very stirring services and whole hearted singing, especially on the last Sunday of term, which was celebrated always as "Sing-up Sunday" and the last hymn of the term was always "Hills of the North Rejoice". The first two lines of the last verse being "Shout while ye journey home; Songs be in every mouth". — usually raised the roof about two inches. We were helped a lot in the encouragement we got from Brookie as he had a host of letters after his name earned by his musical prowess and was a natural on the seat of any organ. We occasionally had the opportunity of hearing him playing in Wantage Parish Church.

Incidentally, the penultimate Sunday of term was known as Ivy Sunday for what reason I know not, but there was always a rush to pick an ivy leaf from somewhere to wear in your buttonhole before going to morning service.

Brookie was a short, almost portly, bachelor with a very kindly father-like nature and, although he had little interest in sport, he made an excellent housemaster for Juniors. He was a stickler for cleanliness and before every meal he stood at the dining room door for "Hands" which had to be shown to him back and front, when he would say either "Pass" or "Wash" before we were allowed to enter. Quite often on Sunday afternoons he would take us all up on to the downs and find a suitable spot, usually on a bank, where we would sit while he read us an instalment from a detective story such as "Raffles" or "Bulldog Drummond" always breaking off at an exciting point to say something like 'that will do for now — we will continue next Sunday and if it's wet it will be in the Big Room' which was our Common Room. Sometimes the suspense was strong enough for us to try and persuade him to continue the story one evening in the middle of the week. With pleas of "Please Sir, come on Sir, please Sir?" We could always tell by the look on his face if our request was going to be granted and I'm sure he loved tantalising us with a drawn-out period of negotiation.

He didn't have a car but owned a very good bicycle with a strong carrier fitted over the rear wheel and he often took the younger members of the house for cycle rides. On one or two occasions he took me home to Sunday tea which I think he enjoyed as much as I did.

The head master Mr K.A.R.Sugden was a tall, gaunt man

with short bristly hair. He walked with a long, quick purposeful stride which earned him the title of "The Portway Express" — I'm not sure if he knew that. Anyway his physique matched his distinctive voice and forthright manner. He tanned me once, giving me four strokes for fighting a day boy who shall be nameless but who had infringed an unwritten rule by uttering a derogatory remark concerning someone's parents, so I knocked his glasses off and thumped him. Naturally he retaliated and we were locked in combat when the Head arrived in the classroom to take us for Latin. Whereupon without questioning I was taken to his study, summarily dealt with and returned to the classroom, where we were all told in no uncertain manner that similar incidents would be dealt with in the same way.

Fortunately a day boy named Evans who had seen it all happen, had the guts to speak up for me, explaining the reason why, which caused the Head to tell my de-spectacled opponent to go to his study at lunch time and he had 'four' as well. We were reasonable friends afterwards, but there was always a something between day boys and boarders and most scraps seemed of the same mix. I'm not really of a fighting nature but had no regrets then or now for what I did, in spite of the unfair advantage of being able to see better than he could once I had knocked his glasses off. I cannot remember any other serious scrap I had at school, other than in the boxing ring but that was different, that was for fun, although sometimes it didn't quite seem like it.

After about a year at Junior Lodge there was a minor alteration in the Junior accommodation arrangements. About ten of us younger ones were moved to a house which adjoined the offices of the Auctioneers, Adkin Belcher & Bowen. It had a small walled-in lawn with a mulberry tree in it, the branches of which can still be seen from the west end of the Market Place opposite The King Alfred's Head Hotel. It was known as Junior House while Junior Lodge in Newbury Street had been re-named Thanet House, I presume the alterations were made to accommodate the entry of a few more pupils.

At first Matron 'Parkie' looked after us on her own but was soon succeeded by a Mrs Griffiths as a housekeeper who was joined later by a Junior housemaster Mr Kirby. The house contained two cellars, the largest one we christened Cucumber Cellar because of its relevant smell. It was big enough to kick a tennis ball around in and we often played five-a-side. It also had a pipe coming into it from the street at one end; it passed straight through from wall to wall and was held firm to the base of a chimney breast in the middle by two metal clips, which made the two recesses at either side of the breast an ideal place to hang on to the pipe, for the purpose of a few swings or up and overs. It was considered to be reasonably safe for one of us at a time, but not two. However, with me on one side and "Deedle" Roberts unable to wait until I had finished my bit of fun jumped up on to his end of the pipe, with the result that the clips gave way and we both fell to the floor, breaking the pipe off close to its point of entry. What we didn't know was that it was a gas pipe and not water as we had all thought. Pandemonium broke out, Mrs Griffiths did her best to find some corks to plug the break; eventually a rag was stuffed into the pipe until the gasman arrived to turn it off outside. Meantime there had been a blow back in the adjoining auctioneers offices, causing a typist to faint, and a complete evacuation of the premises.

We all knew this would lead to some form of punishment, probably tanning. The following day the Head paid us a lunchtime visit, gave us a severe talking to, and demanded to know who was swinging on the pipe at the time, "Deedle" and I owned up, whereupon he said he was not going to beat us but we were to write out the first two chapters of St Matthew in double line copperplate writing and we were to take the first chapter to him in a weeks time; which we did, hoping he would say, right that's enough, but he didn't and told us to bring him some more in a another weeks time even if we hadn't finished it all. Having kept us at it for a fortnight he decided that would do and warned us that any similar escapade would have dire consequences.

Four of us occupied a small dormitory on the second floor which had been converted from a couple of attics. It was approached from the top of a narrow staircase along an open topped passageway which reached to the apex of the roof where a sizeable water tank could be seen and also approached by a crude type of wooden ladder. It was in the summer term and, for some reason, permission to take a morning bath was withdrawn. So by way of a kind of protest our dorm which consisted of Hatchard, Eldred, Lansley, who all came from Basingstoke, and myself, decided to climb up one at a time and have a quick cold dip in the header tank, so with a rota of two keeping "cavey" the operation was well under way with Lansley and Eldred having completed, and Hatchard about to submerge when the warning "Cavey-Kirby" was passed to me while waiting my turn at the bottom of the ladder. Quickly and as quietly as possible I passed the warning on, but it was too late; Kirby who had obviously heard something was going on from the floor below, was past Lansley and Eldred, with towels round their waists, and within a yard or two of me in the nude, to shout "What on earth is going on?" whereupon Hatchard who had already pinched his nose and disappeared from vision below the top of the tank, could hold his breath no longer and burst to the surface with the sort of sound which usually indicates that the water is not only cold but very cold. The outcome of all this being that the four of us were required to bend over in our pyjamas that evening before "lights out" for the purpose of receiving three whacks on the backside from one of Kirby's slippers, delivered I'm sure with the minimum of ill-feeling and in my case reciprocated likewise, as I quite liked him. I have an idea he was a hockey blue and left teaching to take the cloth; the last I heard of him was a long time ago when he was Vicar of a Parish somewhere in Devon.

My next move was back to Newbury Street and the renamed

Thanet House under a new housemaster named Morris, who was married; his wife acted as matron. The year spent there was fairly uneventful except for a daft and dangerous idea of a chap named Stacey which might have burnt the house down. He had the notion that what he called 'fire' football would be fun. The idea was to roll up some newspaper into a ball put some string round it, sprinkle a little petrol on it, ignite it from the top of the cast iron stove used to heat the changing room on games afternoons, pick up sides and kick the 'fireball' around the changing room while in our birthday suits! — In fact the game never got underway, due to a minor explosion when Stacey lifted the lid off the stove to ignite the 'ball' and the immediate flame from the vapour blew up in his face, singeing the hair from the front of his head and depriving him of his eyelashes, as well as leaving some nasty blistering on his face. Fortunately, someone chucked a damp towel on the 'ball' which was thrown outside and what could have been a very serious fire was averted. Poor Stacey was the only one to suffer and it was fortunate that he was 'off games' at the time so his clothes gave him some protection. The house was assembled that evening to be told in no uncertain terms of the seriousness of it all. I have an idea all pocket money was stopped for a fortnight and Stacey's enforced confinement to the sick-room and the pain he suffered were considered to be sufficient punishment for his fool-hardiness in suggesting such a dangerous prank. True to school tradition he owned up and accepted responsibility but in hindsight there is no doubt the rest of us were partly to blame as well for aiding and abetting.

It probably appears from what I have written about school up to now that the learning side was of secondary importance to me, I've no doubt at that particular time it was! I was not particularly bright in that respect. Nevertheless, at the risk of being accused of a bit of trumpet blowing, a glance in the book case reveals that in December 1923 Liddiard R.E.J. was presented with the Upper First Form prize by the Headmaster K.A.R.Sugden,M.A. I have an idea my success stemmed from the fact that my promotion to the Lower Second had been purposely delayed, which meant that I was about a year older than the rest of those with whom I was competing, so it was an odds on chance that I might finish in the first three. The prize was Robert Louis Stevenson's — 'Treasure Island' with the school crest and motto "Dominie quis Habitabit" embossed upon it. I can only hope it cheered mother and father up a little as an entry in the private column of father's account book for that year records that on September 21, 1923, he paid the headmaster £31-13-0 for the ensuing autumn term's board and tuition!

In all honesty that is the only prize I have ever won for any form of academic achievement in my life — nothing to be proud of I know — but a fact.

Before leaving Thanet for School House I had become a member of the O.T.C. (Officers Training Corps) as most other boys did at around 12 years of age. It was about then that Sgt Major Harry Glaysher joined the school staff, replacing Sgt Major Roberts as Physical Training Instructor and O/C the O.T.C. Harry Glaysher or "S.M." as he was affectionately known, was destined to become almost a part of the school for at least twenty-five years and I'm sure there are many who benefited a great deal from his down to earth attitude towards life in general, and the tremendous enthusiasm and encouragement he gave to everybody in the gym and on the games field. Its an old cliché I know but if ever anyone knew how to teach someone to play the game, he did. He looked upon us as 'his boys' and I'm sure in return he had the respect of us all.

O.T.C. Parade was every Tuesday morning following school prayers in the main hall. This meant putting on uniform when getting up in the morning and full dress included the wearing of putties. Now it isn't everyone that is able to master the art of rolling putties around ones own legs, with the certain knowledge that they are going to stay in position for long, so new recruits having run downstairs in a hurry, arrived at the breakfast table in various states of dress which had to be tidied up before the parade in the hope that they would not fall down round your ankles at the first command of 'Platoons — Attention'! I quite enjoyed O.T.C. and was made a bugler in the band which consisted of 4 side-drums, a bass drum and 14 buglers. Toby Cook of Faringdon and I had the honour to blow 'Last Post' and 'Reveille' on the steps of the school War Memorial at the Armistice Day Service in 1925. Other O.T.C. activities were marching, drill and rifle cleaning in the armoury where all the equipment was stored. The old armoury, having been bricked up, still stands near the entrance to the Leisure Centre opposite the school in Portway. Every summer term we took part in a Field Day in which the 'white' army opposed the 'red' army, usually somewhere upon the downs with a couple of regular army officers acting as referees. This was popular not only because it was a day off school but we were issued with ten blank cartridges apiece for use when battle commenced, to make it sound real if nothing else. The other highlight was the Annual Grand Parade at which the salute was taken by a high ranking regular army officer. S. M. always made sure plenty of spit and polish went into preparing for the occasion.

My arrival at School House with two or three others from Thanet called for the usual initiation ceremony which consisted of proceeding, hand over hand, across the width of the Junior dormitory hanging on to a round metal bar which served as a brace on each of the roof trusses. While struggling to get from one side to the other the rest of the 'dorm' had the privilege of belting you with knotted school ties, which could be a bit painful even through pyjamas. Completion of the task did of course mean that you were accepted and had become a true member of the School House.

I can't help thinking the move to a much closer proximity of school facilities changed my attitude a little and seemed to lead to an increased loyalty to the school itself or perhaps I was just beginning to grow up a bit. The fact that I had managed to get into one or two junior teams to play other schools such as Douai, Leighton Park, Magdalen College School, etc. also influenced my outlook. Encouraged by the S.M. I had also become quite fond of boxing and he picked me, as the lightest weight in the school eight to represent the school against Magdalen College School, Oxford: I recall the time between seeing my name on the notice board and stepping into the ring to box someone I had never seen was far, far worse than the three, three minute rounds spent inside the ring trying to convince my opponent whose name was Kirk J.A that I knew more about the 'noble art' than he did — In fact I didn't. He beat me on points.

After quite a few further efforts, including Inter-House contests, usually the most exciting of all, for contestants and supporters alike, the only worthwhile trophy I possess is a silver medal awarded to me in 1926 by the well known flat race jockey Freddie Fox; not for winning I hasten to add but he considered I happened to be the best loser during an evening of school finals — bless him. He always took a keen interest in the schools' boxing and gymnastics. Tragically he died in a car crash on a foggy day near Venn Mill on the Wantage/Abingdon road in December 1945 at the age of 57. This tragedy followed not long after he and his wife suffered the great loss of their only son Michael who was also a pupil at K.A.S. and lost his life while flying Blenheims in the R.A.F. during the war.

It is strange how most anecdotes seem quite clear in my mind even to the point of detail, while the actual timing of their occurrence and relevant changes, bearing in mind that it was almost 70 years ago, must be accepted as approximate.

It was around 1926 that the headmaster Mr Sugden retired and The Rev. F.C.Stocks was appointed as Head. He was a great hockey player being a Cambridge blue and, I believe, captained the English XI prior to the first World War. The school playing field was and still is on the opposite side of the road to the school. At that time half of it was fairly level the rest dipped conspicuously away in a southerly direction towards the boundary which was defined by a brook, on the other side of which was a small grass field dipping in a northerly direction, forming a valley at the western end of which was the school rifle range, being 25 yards with 4 target positions. This was under the jurisdiction of S.M. and used in conjunction with the O.T.C. Although the area behind the targets was banked with a large amount of sand and a red flag was hoisted when firing was in progress, I wouldn't have chosen to use the footpath to Letcombe while the flag was flying, as it passed only a matter of yards behind the range and directly across the line of fire. I doubt very much if its use today would be allowed. Anyway fortunately, as far as I know nobody "missed" the target area by that much, so all was well that ended well.

Returning to the topography of the playing field; the new headmaster was keen to see the field levelled and I think he probably visualised annexing the adjoining field, piping the brook and filling in the whole of the small valley to provide a larger and much better field on which to play games. Although he didn't live to see every phase of his scheme completed, he certainly started it and some of the initial labour costs were kept to a minimum by the occasional use of "school labour" to help level the tons of soil brought on to the site, where he was often to be seen wearing galoshes and encouraging those willing to take part. I remember him saying to me . "Now then Liddiard, you are a farmer's son, so you know how to use a spade and push a wheelbarrow don't you?" Looking at the site to-day with the large Leisure Centre built on part of it, I like to think somewhere under the surface there's a few barrow loads of soil put there by me almost seventy years ago.

I suppose my promotion to School House naturally encouraged me to being a bit more adventurous which led four of us to a spot on Windmill Hill one Sunday afternoon to sample the first cigarette of our lives; having rewarded a day-boy to smuggle us a small packet of "Lucky Dream" or "My Darling" which were both scented and didn't make your breath smell as much as straight tobacco, so we were told by a senior who claimed he had been smoking for terms. In fact our smuggler brought us "Mr Darlings" and of course a box of matches. The experiment seemed to create an exciting atmosphere as we all sat in a ring on the grass puffing away. However, it wasn't until 3 or 4 hours later that things went wrong, when two had to leave the "Tin Tabernacle" during evening service to be sick and the other two of us turned a nasty shade of green but just managed to hang on until after the last Amen. After a further half hearted attempt during which we were nearly caught in the act, I never smoked at school again, although I did try a few Woodbines in the harness room at home during holidays which were given to me by father's groom Harry Eltham. On my 16th birthday, I was bribed by mother and an uncle not to smoke until I was 21 and they would give me £20. Which seemed a lot of money so I accepted, kept a clean sheet and collected four lovely "white ones" as fivers were then called.

The school was without a swimming bath, so during the summer term the facilities of the Town's open air pool were made available on a fixed timetable basis, which was strictly adhered to and operated by a caretaker from within a small alcove just inside the entrance. He was a nice old man who we called Dada Arbry. The bath was situated half way down Mill Street, approached by an alley and steps almost next door to the present book shop owned by Millers. Mondays and Fridays were the worst days for swimming because the pool was drained and filled on Sundays; there were no chemicals to keep it sterile, so Monday the water was extra cold and by Friday, which was known as "Pea Soup" day, it was getting a bit cloudy.

The school "bogs" are worthy of mention. They consisted of seating accommodation in a line of eight cubicles with a continuous channel underneath made of approximately ten inch half round glazed pipes, which ran the whole length of the outfit. The channel was flushed automatically when the water in the header tank at one end reached a certain level and by some ingenious device the sluice gate at the other end opened at the same time to allow discharge, closing again in time to retain sufficient water in the channel to assist each succeeding flush. This operation led to someone coming up with the bright idea of making a paper ship out of a piece of foolscap rather like making a paper cap and turning it upside down, then packing it with some newspaper, installing oneself in cubicle No 1 making sure the other cubicles were well occupied and at the moment the flush was about to finish, lighting the paper and floating the "ship" down stream. With a bit of luck and the timing of launch was spot on, the "ship" would reach No 8 just prior to the sluice gate closing and would be propelled back again to No 1 on the wave created by water as it struck the closed gate. It sounds a bit crude but it worked. Incidentally No 8 was unofficially reserved for prefects and juniors were sometimes sent in for the purpose of warming the seat. The design of the set-up led to real drama on one occasion when a day boy was being tantalised by someone who eventually snatched his satchel from him, ran into the bogs and stuffed it through the hole in one of the seats into the channel below, seconds before the flush was due to operate, whereupon it was carried away only to become entangled in the sluice gate mechanism which prevented it from opening during the following flushes with disastrous results. An immediate school assembly failed to produce the culprit whereupon the Head announced that all exeats and school games would be cancelled until the person responsible owned up. There were only two or three who witnessed the incident and in true tradition were not prepared to "split", which meant it was almost a week before the matter was resolved, during which time the whole school was agog with all sorts of stories and rumours.

By now I was nearly 14 years old and Father and Mother had decided if I was to become a farmer it would be an advantage to send me to Dauntsey's School near Devizes in Wiltshire which at that time was known as Dauntsey Agricultural School and was under the headmastership of the up and coming G.W.Olive, whose contribution toward Education was later awarded with the OBE and who on retirement wrote of Dauntseys in "A School Adventure". A friend of mother's knew him while he was at Cambridge just before the 1914/18 war, where due to his initials G.W.O. he was nick-named Gee-Woo. At school he was known as "Streak" because of the speed with which he traversed the schools long tiled corridors clad in crepe soled sneakers, enabling him to open doors without warning for the purpose of making sure all was as it should be.

Dauntsey's, as it later became, had Veterinary and soil science as well as other agricultural subjects in its curriculum. These became available on leaving the middle form named "Shell". It was from this point that a decision was necessary as to whether to proceed to form 4A with its agricultural bias or into 4E, which was directed towards the more general line of education and the taking of School Certificate and possibly High School Cert.

It was in September 1926 that I entered Dauntsey's as a boarder, accompanied by a cousin of mine of similar age named Norman Wheeler for whom it was a first experience as a boarder. His younger brother Peter entered three or four years later as did Norman's two sons round about 1960. In fact the Vale of the White Horse boasts quite a few Old Dauntsians, father's cousin Charlie Chillingworth (of Prime Dutch fame) father's younger brother Leslie, and a very great friend of mine Victor Tytherleigh who left school in 1917 and I met for the first time when he took the tenancy of Northfield Farm near Faringdon before marrying Mary Reading, a close relation of the Langford family of Readings, in 1938. Sadly Vic died recently and I miss him dearly. We enjoyed one another's company for a long time, having spent happy hours trying to pick winners at National Hunt and Point-to-Point meetings, as well as spending a few holidays together with our respective spouses especially a memorable week in Switzerland. How we laughed many times after recalling a visit to a public park in Lucerne when we stood next to a man at the edge of a pool, watching the water from a huge fountain throwing its jet high into the air. Suddenly he took out his false teeth and proceeded to wash them in the falling spray, put them back into his mouth and walked off as if it was a daily routine — perhaps it was.

Other O.D.'s in the is area were Hamish White, of Uffington, George Twine of Coleshill, Colin Nash of Kingston Lisle, Dick Roadnight of Brightwell near Wallingford, Ted Gent of Kidlington all of whom were farmers. A wider spread contained the names of A.G.Street, the farmer/novelist who wrote Farmers Glory, Strawberry Roan, Wessex Wins and others. He farmed close to Stonehenge and for many years wrote weekly articles on the farming scene for the "Farmers Weekly". Perhaps the best known of all and possibly became one of the most wealthy was the Rev. W. V. Awdry of "Thomas the Tank Engine" fame.

He entered Dauntseys as a boarder in 1924 a year before I did. The school was then officially Dauntseys Agricultural School. It is interesting to know that his family had connections with the school for a number of years, his uncle Charles Awdry (1847-1912) and cousin Col. R. L. Awdry (1881-1949) both having served as Chairman of Dauntseys governors in 1911-12 and 1937-49 respectively. My recollections of W. V. Awdry are that he was a quiet, studious, lanky type who played cricket a few times for the first XI.

Another was novelist Nigel Balchin who will be remembered by many for his book "The Back-room Boys". He was Head-boy when I arrived and his parents owned the village shop where we bought tins of peaches, apricots and sardines as well as doughnuts which were all helped down with Salisbury beer on last night of term dormitory feasts.

Among the names of School Governors at the time I remember Lord Baden-Powell. Later the chair of this august body was occupied by Marshal of the R.A.F. Lord Tedder whose son was a contemporary, nicknamed "Mousie". We were sometimes called upon to address envelopes to parents containing Speech Day invitations etc., and I recall with pride having addressed one to his father, then a Squadron-Leader.

The School has strong links with the Mercers Company and by a strange coincidence the last vicar to be appointed for our parish here in Baulking, before we became grouped with Uffington and Woolstone, was The Reverend F.P.Harton who was also a member and past Master of the Mercers Company and was appointed by them to serve on the Governing Board of Dauntsey's.

The school numbered around 110 pupils about 80 of which were boarders. We were divided into three houses, Dauntsey, whose members of which I was one, wore black ties with red diagonal stripes, Mercers tie had green stripes and Fitzmaurice blue. The only time we assembled in houses was at meal times when each house had its own tables in the Dining Hall, on the walls of which were displayed team photographs of the past, school trophies and house cups.

The staff dined at a separate table in the middle of the hall and those I remember are the second master E.R.B.Reynolds, "Erbie", E.L.Batten, R.S.Barron, "Georgie", J.A.Davidson, "Jad", — Harrison "Tweedie", W.S.Segger "Willie", A.L.Coates who was in charge of the farm small-holding and the school rugby XV, and F.N.S.Creek the great amateur footballer who gained a blue in his first term at Cambridge and played in the first University soccer match after the 1914-18 war on the Queen's Club pitch in West London, which he described as being like a bowling green and where rows of chairs were drawn up a few yards from the touchline, which were reserved for lady onlookers, their escorts having to stand behind for the best part of a couple of hours. The playing kit consisted of thick flannel shirts without numbers, — they came in following the Second World War — and long baggy shorts with essential pockets for handkerchiefs and peppermints. They also wore thick stockings with shin pads underneath and a pair of heavy boots. The footballs themselves became increasingly heavy during the game, especially in the second half if the ground was wet, and it required considerable skill — and courage — to head a goalkeepers clearance kick. Nevertheless the standard of play must have been relatively high, because not only did Oxford & Cambridge provide the backbone of the great Corinthians side of the 1920's and early 30's but their blues obtained many amateur international caps and a select few of which Norman Creek was one — were picked for full international honours. In retirement he continued to do much for the future of football and covered many matches with live commentaries for B.B.C.radio. "Jad" Davidson was our very keen music master and aspired to putting on Handel's Messiah in Devizes Town Hall for a couple of evenings in the spring term of 1928, for which the school orchestra was strengthened with local talent and the choral society likewise, with the addition of two or three professional soloists, one being the well known soprano Wyn Ajello. I remember this for the reason that having taken part in rehearsals almost up to the evening of presentation, my voice broke and "Jad" told me I could continue to appear in the choir but on no account was I to make a serious attempt to sing, other than perhaps in the Hallelujah chorus which he was sure would drown any peculiar sounds I might be likely to make. I think the presentation was considered to be a great success and I recall Matron Acton, who was a kind rosy faced lady of ample proportions, saying with tears in her eyes that the rendering of "I know that my Redeemer liveth" sung by Wyn Ajello was something she would remember for the rest of her life. Mention of matron reminds me of "Sanny Nurse" to whom matron passed us on should we need a few days nursing in the Sanatorium which had about six beds and stood in a slightly isolated spot within the school grounds. Fortunately, I only spent about three days under her care with a touch of 'flu and apart from not feeling well, somehow we didn't seem to get on very well together. I was informed later that she could be a bit moody and was probably still suffering the effects of a row with one or two other boys who had preceded me and disposed of a few portions of rice pudding in the sick bay ventilators where it had remained undetected until the aroma revealed its whereabouts.

There being no school chapel, Church services were attended at West Lavington and Little Cheverell the latter often being taken by Erbie Reynolds or Creek as lay readers, with prefects sharing reading of the lessons with local residents such as Colonel Awdry and Admiral Luce.

The area provided ample scope for Sunday afternoon walks, and the natural bathing pool close to the railway viaduct was often frequented and there was always the possibility of coming across a wild duck's nest with an egg or two to tempt the finder. I have a photo taken of me by the side of the brook which fed the bathing pool, with the viaduct visible in the background. It was taken by A.M.I.Austin who went into the Colonial Legal Service as a barrister in Malaya, Malta and Guyana.

At the rear of the school kitchen area there was a small yard in which stood a dozen or so dustbins attracting considerable attention from the local population of starlings and I remember a school prefect G.A.Ransome asking one or two of us if we could catch a few birds for him to carry out some vivisection experiments in the lab. So all the bin lids were put on tight except for one which was half full of food scraps. The lid to this one was raised to an angle of about 45 degrees and propped open by a small forked stick to which was attached about 25 yards of string to allow the trap to be operated from a hide so that the starlings approach could be observed. The method was simple, just wait until 3 or 4 were inside and pull the string. To the best of my recollection the success rate was around 50%; more than enough to satisfy Ransome's experimental needs. He was very keen on the science of life in its various forms and I remember watching him dissect a grass snake with a half digested rat in its gut. The operation was performed on the lid of his tuck-box and attracted quite a gallery of onlookers. All this must have stood him in good stead as he later entered medical school and was eventually knighted Sir George Ransome for his contributions in the field of medicine.

At the end of what was his last term at Dauntsey's he was involved — as I was, to a much lesser degree — in a bit of drama. It was end of term concert evening and an O.D. named Parkinson had arrived at school on a sporty looking motor-bike. It was a Francis Barnett 2 stroke. R.J.Carruthers a fellow prefect of Ransome's, also due to leave at the end of term, chatted up the motor-bike's owner prior to the concert for the purpose of taking the machine for a spin. With the owner's permission, during the concert interval Carruthers took off on this somewhat powerful machine towards Earlstoke two or three miles away, along a road, that has two S bends, one incorporating quite a dip which that evening contained some low lying mist, with the result that Carruthers and bike crashed through the hedge into a field, where they lay for a few minutes before Carruthers was able to drag himself back on to the road.

At the end of the concert while making my way across the yard in partial darkness to the "bogs" I was shouted at by one of two people standing in the shadows about a dozen yards away. I was told to "stop", and not to come any closer but go and find Ransome as quickly as possible and tell him to come to the "bogs". Recognising the voice to be that of Carruthers I did as I was told. Being the end of term there was little time for the story to unfold but Carruthers, having damaged the bike and worst of all himself, wanted Ransome's opinion as to how serious his injuries were and if he could possibly conceal them until it was time to leave for home next day. The first few days of the following term failed to reveal the outcome of the story except that Ransome had quickly expressed his opinion that Carruthers needed immediate treatment for his injuries so despite everything he was carted off to the San. The other person with Carruthers on the night of the incident turned out to be a young local Vicar who picked him up and brought him back to school in his car having been persuaded not to hand him over to the headmaster until he had seen Ransome.

During my second year I was appointed post boy, with the duty of clearing the school post box after breakfast then calling at the Headmaster's house half way down the school drive to pick up any further mail, which I then delivered to the Littleton

Pannell post office cum village shop, situated close to the school entrance and near to The Wheatsheaf public house, which on occasions it was necessary for me to visit via the back door for the purpose of delivering small amounts of money wrapped in bits of paper bearing a horse's name and betting instructions, given to me by one or two seniors who liked a gamble. This was the easiest way in which they could get their money on, after a look at the communal morning papers displayed in the entrance hall. I was always relieved to get back into the school drive, which was flanked by two long rows of lime trees, as all pubs were strictly out of bounds. Winnings, if there were any, were not my concern and were collected from the landlord by the punters.

The small post office and shop was run by Reg Hillier and was not out of bounds. He had a daughter who occasionally helped behind the counter which attracted visits from some of the school 'elders', one of whom was G.(Guy) Middleton. On one such visit; Middleton noticed a small green bus owned locally by a Mr Sayers, used to transport people and parcels etc. from around the villages, to Devizes, about 5 miles distant. It was standing near to the front of Reg Hillier's shop which had a small forecourt divided from the road by iron railings and a gate. Middleton also noticed as he approached the gate a length of chain with a hook on it, dangling from the parcel carrier at the rear of the bus. Turning to close the gate behind him, with slight of hand he hooked the chain to the railings, just as the bus moved off and then ran into the shop to tell Reg Hillier he didn't know what was happening outside but his iron fence seemed to be disappearing down the road. Guy Middleton became an actor, appearing on the stage and in supporting roles in quite a few films. I saw him in a thriller named "The Crooked Billet" at the Empire Theatre in Swindon, before it became a cinema and was eventually demolished to make way for development in the area of the multi-storey car park and the precinct of the present Wyvern Theatre.

It was now 1928. Farming as well as every other industry

was still in deep recession and struggling to find its feet after the General Strike of 1926 which virtually brought the country to a standstill for 3 or 4 days. What services there were depended on the loyalty of a few, backed up by some of those thrown out of work but were willing to help in jobs they knew little about. As in the case of two men in mufti struggling in the crowd to get on and off an underground train, the one in front saying "I must get off, I'm the platform porter", and the other answering "I'm sorry ol' boy, I've just been to the Gents, I must get back on I'm the driver".

I was not yet 16 when father and mother decided they could not afford to keep me at Dauntsey's for a third year. I remember them taking me out to lunch at The Castle Hotel, Devizes one exeat Sunday to explain the position and how serious the state of farming was. So much so that father was considering returning to the grocery trade and had already registered an interest in a branch shop belonging to one of his old business rivals Messrs Rant & Tombs of Abingdon who had decided to put their shop at Stanford in the Vale on the market. However, it was thought that my home coming would not only save the £38 per term school fees but benefit labour costs at home in reducing the number of paid staff by one.

Further proof of the prevailing conditions can be drawn from the letter *(reproduced below)* written to father by Headmaster Olive in which he pleads on my behalf.

(From) *Headmaster: Geo. W. Olive, M.A.*
Dauntsey School,
West Lavington,
Near Devizes,
Wiltshire.
July 5. 1928

Dear Mr. Liddiard,
  I want to tell you what is passing through my mind with regard to your boy.

> *I was sorry when you told me at the beginning of term that your boy would be leaving, because it seemed to me that he was developing from boy into man in just the way, I know, both you and I want to see. I am even more sorry now, because I do feel—and other masters, who think about these things, feel the same, that he is developing in every way that matters, just as a boy should develop—and a term or so more at school would mean all the difference to him.*
>
> *Let me be quite frank. I am not writing like this in order to keep him because of the vacancy he creates. I have more boys wanting to enter the school next term than I know what to do with.*
>
> *I know also that everything connected with agriculture is about as bad as it can be, that parents who are farmers have to look at every penny. — But I have thought it carefully over, and I feel that I would rather finish the formation of character in your boy than take a new boy in his place.*
>
> *If you cannot let him return, I shall know that it will not be for want of consideration on your part, and at the same time I shall have done what I feel is my duty.*
>
> *I hope you enjoyed Speech Day Week End.*
>
> > *Yours sincerely,*
> > (Signed) *Geo. W. Olive.*

There is no doubt that I was pleased to leave at the time but, ignoring the financial side of the issue, I'm sure in retrospect the Head was right and it is possible a final polish may have helped, but who knows. The two things I missed out on most, were that I was unable to proceed from Shell form into 4A and so missed the opportunity to put some of the theory already acquired into practice on the school smallholding. The other disappointment concerned the possibility I may have had in furthering my performance on the games field. I had progressed through Junior and Senior Colts to the odd game or two in 2nd

XI and XV in cricket and rugby and have a feeling I might just have made the first in both — but alas it was not to be. The schools I remember playing against most were Monkton Combe and Downside. It was an added treat to be picked to play at the latter as we were always entertained afterwards in their indoor swimming pool. I have a recollection that, being a Catholic School, cricket games were interrupted for a few minutes when the Abbey bell sounded for evening Vespers.

The only tangibles I have to remind me of Dauntseys and which decorate the mantelpiece in our sitting room are two copper shields mounted on oak, displaying the school crest and underneath the fact that I was second in the 120 yards hurdles in two successive years. I take pride in the fact that they were once observed by Lord Burghley when he called at Church Farm in his capacity as Joint Master of the Old Berks Hunt to inform us that hounds might possibly be crossing our land, following a nearby meet and would it be alright for them to do so. His signature in our visitors book with that of Mrs. Doris Bean, the other Joint Master, confirms that the visit was on February 4th 1954. Many people will know that Lord Burghley was awarded a Gold Medal for winning the 120 yards hurdle race in a world record time at the 1928 Olympics held in Amsterdam. It was a record he held for a few months before it was, I believe, taken from him by an American. Holding one of the shields in his hand he said "I see you were second; funny thing I was no good at school, it wasn't until I got to Cambridge and Fenners that I got the hang of it."

FIVE
Growing Up

Having left school in September 1928, life began to take on a more serious aspect, It was usual for farmers' sons to follow in their fathers footsteps. In a lot of cases landlords were quite willing to transfer tenancies to partnerships formed when fathers considered sons were old enough to join the business. As I was only 15½, time for this was far from ripe although such a move would have been relatively easy as Grandfather had recently made the farm over to my father, who up to then had been paying the annual rent of £275. However it was decided that I should spend a year or two working on the farm for my keep and a starting 'wage' of seven shillings and sixpence a week, half-a-crown of which I gave to mother as a token of recognition that although I was being kept, I should be reminded that money didn't grow on trees and I shouldn't expect something for nothing.

Due to the depressed state of farming there was also considerable doubt as to whether it was wise to stay in it, so it was understood that this was to be very much a trial period. Father suggested I had a small farming enterprise of my own, for the purpose of experiencing the feeding and caring of something which was my own and hopefully would be "a nice little earner" for me as Arthur Daly would now say, enabling me to save or spend a little as I wished. As there were four pigsties at Church Farm, I could use, he set me off with half a dozen 7-week old weaners, just off the sow. I see from my first account book they cost father £7 on October 8th 1928 — consumed 12cwts of toppings (a wheat by-product) which cost

£4-6-11 and I sold them as porkers at Faringdon Market for £14-5-0 showing a profit of £2-18-1. It is interesting to note also that the price of toppings on October 8th 1928 was £10 per ton and by January 5th 1929, it had fallen to £7 per ton, due to tumbling corn prices.

I borrowed £4 from father to put with the £14-5-0 I got for the porkers and made the first deal of my life, buying a crossbreed sow and her litter of twelve from old Mr. Brewer, who rented Spanswick Farm, Letcombe Bassett, from Grandfather for £128 per annum, the equivalent of 50 pence per acre. In those days the farm grew a small acreage of oats and barley while the remainder consisted of downland grazing, some of which was contained in a 60 acre penning on a north facing bank which Dick Brewer sometimes let to father for summer grazing. Providing it was not overstocked, young cattle did well on it from about the middle of May onwards. Dick used to say "Tis no good sending 'em up before then 'cos it takes time to warm up, up 'ere".

After seven weeks I selected three gilts of the litter for breeding, sold the remaining nine pigs and took the sow to Mr. John Gillings's Wessex Saddleback boar at Fernham, the service fee being 5 shillings. I reckoned then that my pig enterprise was well on the way and that the four sties available would soon be full.

During this time people were beginning to take an increasing interest in the wireless, the B.B.C. already broadcasting regularly from their London Station 2LO on shortwave and Daventry 5GB long wave. The Wheeler family at the brickyard were the first to own a receiver in Baulking and I had been invited to "listen-in" as it was called, on quite a few occasions. Their youngest son Johnnie seemed to have equipped himself with a fair knowledge of how receiving sets worked and although his father was a builder, Johnnie, unlike his older brother Bill, was far more interested in the mechanical side of things, such as engines, motor cars and wireless sets. Their set was a Marconi three valve; adaptable for either loud speaker or head phones. Quality of reception varied a great deal and at times atmospherics were so bad that programmes became almost inaudible. This was thought to be due to interference from the Government wireless station situated high up on the Cotswolds at Leafield, a small village about 3 miles north of Burford. It was quite a landmark and its eight tall main masts could be seen on a clear day from at least ten miles away. I think it was used a great deal during the First World War for the sending and receiving of important coded messages tapped out in morse. It was completely dismantled soon after the Second World War and the site returned to agricultural use. Although Johnnie Wheeler was a few years older than me I saw quite a bit of him and he told me the basic requirements necessary to extract sound waves from the ether, in the form of a crystal set with a pair of headphones and a couple of copper coils which could be wound rather like knitting, on a coil maker; the completed coil being released by the removal of all the needles from a central wooden disc into which they were slotted before winding commenced. In fact he installed the first wireless set we had at Church Farm. It was a crystal set which he made himself, the circuits and coils being fixed to the underside of a nine inch square of ebonite fitted to the top of a neat wooden box. The crystal and catswisker, which needed adjustment to obtain the best reception, together with terminals for aerial, earth and headphone connections were mounted in the topside of the ebonite. The outside aerial was about twenty yards in length being attached to a 16ft high scaffold pole at the end of the garden and the roof eaves at the other, the lead-in and earth wires entering the house through the corners of the sitting room window frame. With a lightening breaker switch inserted into the earth wire and fixed on the inside of the window frame; this needed to be switched to the safety position before retiring to bed every night in case of thunderstorms and lightening. Having watched and helped Johnnie with the installation I felt competent enough to persuade a few family friends to allow me to install similar sets for them the price of which varied from thirty two shillings (£1.55) to £2 depending on whether they required one or two pairs of headphones. This may not sound a lot but related to the basic agricultural wage of 28 shillings a week puts it into perspective. It certainly provided a few more shillings for the coffer and no doubt helped towards buying a few things for myself such as the first suit of clothes I bought on 1st December 1928 for £2.19.6.

As the months went by farming became more and more difficult as prices dropped. Quite a few of the village lads in their late teens left the land to join the Army, it being about the only way in which they could broaden their horizon. There were few jobs available within the countryside itself and lack of transport made it difficult to get to work in the big towns, although one or two people from villages within cycling distance of Uffington Station were beginning to catch the early train to Swindon where they had been lucky enough to get a job but it was a struggle to compete with the town workers living on the spot.

With the knowledge of these circumstances and listening to advice from friends, usually while playing a game of "ha'penny nap", I decided it might be a good idea to find out what the prospects were in a business, concerned with some kind of mechanical engineering, in the hope that someone might take me on as an apprentice and teach me a trade. The R.A.F. was an early consideration but I was not particularly attracted by the possibility of a service life probably because I had already been looked after too well at home. A long and helpful talk with Frank Lane, "Uncle Frank" to me, who owned an agricultural engineering and ironmongery business in Faringdon and who with his wife "Auntie Blanche" were lifetime friends of the Liddiard family, resulted in an interview with Mr Skurry a founder of the well known garage of that name in the Old Town district of Swindon. Following a brief look round the garage workshop and longer recitation of the difficulties existing in the motor trade, he finished by saying that, although he would like to have a member of the Liddiard stock as an apprentice, he felt it would be difficult for me being twelve miles away and he favoured taking lads from the town as they were more likely to stay on with him after serving their apprenticeship.

My next visit was to Stuart Turner's, the marine engineers at Henley-on-Thames, who were a much larger concern employing a sizeable work force. Following an interview I was offered an apprenticeship but despite the efforts of some elderly friends of my mothers who lived in Henley they were unable to find lodgings for me, so I had to turn it down.

It so happened that Tommy Anns had recently sold his ironmongery business in London Street, Faringdon, to a young man named Percy Thomas who was quite a live wire and soon integrated himself into the business life of the town. Conversation following a deal father had with him for a lawn mower led to the fact that he was prepared to take on an apprentice to join his staff of three and if I would like to work for him for a trial period of 6 months to enable me to come to a decision as to whether this was the kind of business I wanted to be in and if he considered I was suitable, then we could discuss the future from there. Although this was not quite the experience I was looking for, it had connections with my leaning towards the mechanical side of business, so I accepted Percy Thomas' offer round about my seventeenth birthday and spent the following 6 months working for him, cycling the five miles to and from Faringdon, all for the princely sum of seven shillings and six pence per week. I was very fortunate in being provided with a mid-day meal by my Uncle Pete and Auntie Peggy who had recently married and lived at North House next door to Romney House where Grampy and Granny lived in Gloucester Street.

It wasn't long before I realised this was not the kind of career I was looking for but I stayed the 6 months probation period and gleaned a few useful insights into running a retail business. Percy Thomas impressed me as a keen, enthusiastic, but careful man. For example, he told me his business had increased and would benefit from the acquisition of a delivery van, but he hadn't got the money to buy one, so he got two of us under his guidance to remove the rear part of the body of his Morris Cowley two seater motor car by hack sawing it off at the point where the doors closed and bolting to the chassis in its place a covered box van body with a tail board, which we made mostly out of ply wood. This allowed fair sized pieces of furniture etc. to be transported during the week; while converted back to the original it was available for pleasure at weekends. I recall being fascinated by Jack Berry the local sign writer when watching him decorate each side of the van with Percys' trade description and telephone number etc. the experience of being involved in this dual purpose enterprise, small though it perhaps seemed at the time helped me to realise there were other ways of reaching a goal other than by spending hard cash. Grampy Ernest underlined this to me a year or so later by saying "Always remember my boy, when you are contemplating a purchase; it isn't always what you can do with, it is what you can do without".

With my short taste of what it might be like being a tradesman over, it was make your mind up time and after a great deal of talking it over with mother and father who made it quite clear that farming would not be easy but if I decided to stay at home father would take me into partnership in a year or so and as the farm was now his, due to Grampy's recent generosity, if I was prepared to help all I could and behave reasonably, being an only one, it would probably be mine another day. Considering that to be a reward well worth struggling for and having already spent ten happy years living at Church Farm, come what may I would endeavour to make it many more.

Although still very apprehensive about the future of farming I am sure mother and father were pleased with my decision and remember them reminding me that despite a great deal of effort the elements could be very cruel at times and that there would be disappointments. Mother summed it up by saying "You can go to bed sometimes, quite pleased with yourself and thinking everything is rosy and wake up in the morning to find you are not so clever as you thought".

So it was that having prepared for what may lie ahead, 1930 saw me take the first real step on the farming ladder which was to end with my retirement some 50 years later.

On the assumption that "All work and no play makes Jack a dull boy" mother and father accepted my need for a certain amount of relaxation — provided there was not an over amount of emphasis on the play side of it. Being a farmer needs a certain amount of self-discipline and the farm, can soon miss the benefit of the guvner's boot. Most farms sported a tennis court, some better than others. The amount of care and preparation they received, resulted in the amount of accuracy and height to which a ball might rise but it was the same for everyone where-ever one was playing and as one of my contemporaries once said years later — We had a lot of fun with those slack rackets and soft balls. But it wasn't all like that, there was good tennis played at The Grounds Farm, and The Manor Courts, Uffington as well as at Knighton, Fernham, Clanfield and elsewhere. Most parties finished up with sit-down suppers and a good old sing-song around the piano, especially at The Manor, Uffington where the McIver family lived and at Knighton where their relative Ken Fraser — Uncle Ken to all of us — lived and farmed.

John McIver moved down from Scotland to become an agent for Lord Craven on the part of his estate situated in and around Uffington and Ashbury which included Ashdown House, built by Lord Craven in 1665, in an isolated part of the Berkshire Downs in the desperate hope that he and his family would escape death from the Great Plague which was rife in many parts of England at that time. "Uncle Ken", a bachelor, looked after by his sister Meg, was a brother-in-law of John McIver and for a few years during summer holidays they were visited by members of the Kidd family, some friends of theirs near

Edinburgh. They were a family of five boys and four girls and when staying down here added greatly to the party scene, joining in the tennis and occasional cray-fishing parties under the light of the harvest moon with great gusto. The enjoyment of the sing-songs owed a great deal to Stella, John McIver's daughter who accompanied us on the piano in such renderings as My Blue Heaven, Ramona, Only a Rose, Ol' Man River, Rose Marie, Why do I love You, One Alone and many more.

Stella married one of the Kidd Family, Jimmy, just before the Second World War in which he served in North Africa and was a member of the garrison which held out for so long against Rommel in the historic Seige of Tobruk. After the war he returned to his Edinburgh firm of solicitors but sadly died at too early an age. He was a great fellow, his elder sister Margaret took silk and had the distinction of becoming the first practising lady barrister in Gt Britain.

Another summer relaxation was Saturday afternoon cricket. Grampy was quite an enthusiast and told me he played for Faringdon in the 1880's when the ground they used was near the foot of Jasper's Hill close to where the O.B.H.Kennels are today. He was in fact a member of the Kent County Cricket Club for a number of years and rarely missed Canterbury Cricket week, staying with his daughter Edie, who was married to Hubert Holtom, one of the last generation of flour millers of that name who operated a sizeable mill at Ducklington near Witney, now worked I believe by Oldacres. It was usual that the cricket week coincided with the Kent Lamb Sales at Ashford so it was opportune to combine a little business with pleasure and if the price was right a few lambs were transported to Steeds Farm, Faringdon for the purpose of tidying up the late summer and autumn grass and hopefully to make the trip self-supporting. Although he was getting older he was still keen to hunt the nimble shilling.

Anyway his liking for cricket prompted him to offer to pay my subscription of £1 (there was a special rate for younger players) so that I could be a member of the Faringdon & District

C.C. which was then composed mainly of well established farmers, such as Guy Weaving, Victor Arkell, Arly Barton, Bob Weaving, Tom Gauntlett, Ken Nibbs, Victor Adams all under the captaincy of Stanley Adams. The Club must have been in existence for more than a hundred years as I had the honour to attend the Centenary Celebration of the Abingdon Cricket Club at which I was shown a letter from their archives confirming arrangements for their first match which was to be played against Faringdon.

After a couple of seasons, the Hon. Sec. Sid Luker very tactfully suggested I should now consider paying the full subscription which was £3. I didn't like mentioning this to Grampy; so I rustled the extra couple of quid from somewhere and said I was sorry but would not be able to play next season as I really couldn't afford it and, kind as the older members were, I felt unable to stand my corner which was unfair on them during after-match hospitality often dispensed in the nearest pub to the ground. Club bars in the pavilion didn't arrive on the scene until after the second World War. Quite a few Hunts ran a cricket side in those days too. Play was by invitation and games were one day matches, which usually took place on privately owned cricket grounds attached to country houses in which both teams were entertained to lunch, tea and after-match drinks. I recall being lucky enough to play at Bruern Abbey the home then of Mr. Crompton-Woods, also Aldermaston Park (The Miss Keysers) Stanway Park (Sir Bolton-Eyres-Monsell) Englefield Park (H.A.Benyon Esq.), also at Poulton Priory the home of Col. Mitchell whose family suffered so tragically while attending Sunday morning service in the Guards Chapel, Kensington Barracks, when it received a direct hit from a flying bomb during the Second World War.

A neighbour, Sidney Reade who farmed at Manor Farm, Baulking, whose family consisted of two sons, the elder Maurice, who married a cousin of mine, Jim, who was destined to breed the memorable hunter chaser, 'Baulking Green' and two daughters, Mary & Ethel. They were all a little older than me but we visited one anothers homes frequently and I spent many happy times with all of them over the years. Maurice who was playing cricket regularly for Uffington suggested as I couldn't afford to play for Faringdon, why didn't I join him at Uffington, where the subscription was only five shillings. The pitches wouldn't be so good but they had a lot of fun on Saturday afternoons — Sunday fixtures weren't even talked of then. I liked the idea, was accepted and we played in what is now the recreation ground. The square being roughly 100 yards south-east of where the Thomas Hughes Memorial Hall stands today, the only part of the field which was 'mown' and was protected from grazing stock by a temporary wire fence when not played upon. The outfield was trimmed occasionally with a two horse 4ft six finger bar mower used for haymaking. Unlike any other field I have ever played on it had what looked like a cluster of small pits left behind after some kind of soil excavation. This unusual feature provided anyone fielding at square-leg or extra-cover, (depending from which end the bowler was operating) quite a different view of the game as he could be anything up to five feet below the batsman, making the chance of catching someone out quite an interesting exercise.

This is akin to a friend, Brian Bowden, telling me more recently that while on a club cricket tour in the West Indies he had the good fortune to have a chat at a party with Wes Hall, the famous W.I. fast bowler of the 1970's. Wes asked Brian some of the places where they had played, and recalled that he had played in a charity match at one of them. He reminded Brian that due to the ground sloping it was impossible for the batsman at one end to see where the boundary was at the bowlers end and as he needed to start his long run from the boundaries edge he couldn't be seen either. So the umpire having given the batsman guard, signalled to Wes to start his run, at the same time informing the batsman that "He's a coming" underlining his warning a few seconds later by saying "He sure IS a coming".

Completing the description of the Uffington ground, there were no boundaries, other than six into the road in front of the council houses, so it was hit as hard as you could and run as many as you can. The pavilion was about the size of a small hen-house and was situated under an old withy tree near the road.

The captain when I joined was Reg Matthews, a member of the large family of Matthews who have farmed at Spencers Farm, Baulking for many years and whose nephew Jim still does. Reg was a good slow left arm bowler and story has it that in his younger days he won a match against Stanford in the Vale almost single handed.

Apparently Uffington batted first and were all out for 28 runs, in reply Stanford had made 27 runs for four wickets when Reg took five wickets in six balls to make the score 27 for 9, Joe Bailey then bowled a maiden over for Uffington and Reg took the remaining wicket with the first ball of his next over, giving Uffington victory by one run. What made the performance even more extra-ordinary was that all six batsmen were "bowled", and the ball was presented to Reg after the match, suitably inscribed with the particulars of his memorable feat. In fact Jim Matthews still has the ball.

Another Baulking farmer, Percy Reade of Colliers Farm, having played for Uffington in the dim past was now the Clubs 'official' umpire and even in the hottest summer weather he carried out his duties as such clad in breeches and well polished brown boots and gaiters, wearing the traditional white coat and always a black bowler hat. Unfortunately, when excited he was occasionally affected by a stammer, which caused him to be known by the younger generation, rather disrespectfully as "Blink-Bonnie". During one match there was an almighty appeal for a very close "run out". The batsman, being a true sportsman and thinking he was out, didn't wait to see if Percy had his finger up, made his way back to the pavilion only to be recalled by the fielders shouting that the umpire had only just managed to say "Not Out". So he went back and had another go.

I remember Sidney Reade teaching me to skate. It was about 1926/7, all the ponds were frozen over and the one on the village green in front of their farmhouse was providing great fun for sliding. An old cupboard in their rambling old farmhouse produced a few pairs of somewhat ancient skates which we attached to our boots by screw clamps which needed a bit of de-rusting and oiling before they would work; others were secured by leather straps and string. Supported by an old strong kitchen chair I ventured gingerly on to the ice followed by Sidney who promptly pushed me in the middle of the back with his long thumb stick, with the result that the chair went high into the air and I finished well and truly on my backside. "Get up" he laughed, "You'll soon learn how to stand". As soon as I was on my feet he pushed me over again.

After a few repeats I managed to scramble away from him and remain upright, whereupon he said "Now you'll soon be able to strike out on your own, without the chair" — that's how I was taught.

I suppose I turned out to be just about adequate on skates and was able to join in the fun on the few occasions when the lakes froze over at Faringdon House and Buscot Park, and the ice was considered safe enough to allow the public to skate on them. I remember one afternoon at Buscot when there must have been at least fifty or sixty people on the ice, it was suggested everyone gathered together to discuss organising a game of ice-hockey the following afternoon, providing a thaw had not set in and anyone interested would they bring a hockey stick or some sort of stick to play with.

The next afternoon when most people were limbering up someone blew a whistle and summoned us all to a point somewhere near the middle of the lake in order to pick up sides. All was well until the concentration of people in the one spot built up to about three and a half tons causing the ice to let out three or four ear piercing sounds as superficial cracks raced their way to points on the lakeside. I've never seen a small crowd of people disperse so quickly. However, after a certain amount of inspection and jumping up and down, it was decided safe enough to proceed and a rare bit of fun was had by all. The star players I remember were Ivor Cooper, who managed Harris & Matthews Corn Stores at Stanford in the Vale and whose brother Geoff played ice hockey professionally for Bournemouth on Saturdays and Southampton on Sundays. I remember him telling me he was paid £5 for each appearance plus travelling expenses from Oxford where he lived. Others outstanding were George & Bertie Adams of Fernham, the rest of us made up the numbers and probably spent more time on our backsides than our skates. It is possible some of us improved a little when the Ice Rink opened in Botley Road, Oxford. It was quite a new and enterprising venture and provided the opportunity for a very good evening out, not only to skate but to support the Varsity Ice Hockey team who at one time vied with Grosvenor House Canadians for the League Championship. The rink was also used a few times for staging Varsity Boxing matches which about six of us from the Faringdon area rarely missed. Unfortunately, the whole enterprise failed to provide the success hoped for and it was said due mainly to high running costs it had to close down, the premises being sold to Messrs Coopers for marmalade making.

Oxford being only 20 miles from Baulking, the Varsity provided a great deal of sporting entertainment, such as boxing, mostly held at the Town Hall, rugby in Iffley Road, cricket in The Parks, as well as Iffley and I had the pleasure of watching many great players who went on to play for their countries and others who were already internationals and came to represent the All Blacks, The Springboks, The Wallabies, as well as those invited to play for the Major Stanley's XV prior to the Varsity Match at Twickenham every year.

The first cricket match I watched on the Christchurch ground in Iffley Road, was against the Australian side captained by Woodful in the 1930's and from memory contained McCabe,

Ponsford, Bradman, Kippax, Fairfax, Richardson Grimett, Oldfield. The Aussies put together just over 500 runs on the first day, Ponsford making 205 not out and McCabe 98, I think Bradman who was well on his way to 1000 runs in May was bowled by Garland Wells for 38. The thing that impressed me most that day was watching the great Aussie spin bowler C.V.Grimett in the nets before the match. I found it difficult to believe that he could make the ball break a foot or more either way. I have never seen anyone do it to the same extent until witnessing the performance of another Australian, Shane Warne on television some sixty years later.

Winter sport after leaving school started by playing football for Uffington on at least four different fields, one of which was Wentworth's small field on Woolstone Corner (where the battle with the gypsies took place). Being small it meant that the goal at the eastern end was only about fifteen yards from the garden hedge of a pair of thatched chalkstone cottages one of which was occupied by a great old character named "Bristle" Breakspear (his hair was short and bristly). He was a forthright old chap and although he would be the last to admit it I think he very much enjoyed watching the game from his garden over the top of his well trimmed hedge. On one rather windy afternoon, following quite a few shots at goal, the ball found its way into Bristle's garden more than usual. Having thrown it back over the hedge each time in his usual somewhat slow and deliberate manner, he exclaimed following a further return that "If you kicks 'ee over yer again, I shall bust the bugger".

It was said that Bristle could neither read nor write; I'm not sure if this was correct, if it was then it seemed to matter very little to him. During the second war he offered to help us with singling and hoeing sugar beet or mangels on an hourly basis. As he was elderly he came and went when he liked, he never carried a watch but he would call in at the farm every morning to ask the time before starting work. On arrival at the root field he would set about half a dozen sticks in the shape of a sundial on a small piece of land close to the field gate. Even if he couldn't read, a quick glance at his sticks when the sun was shining gave him a pretty good idea when it was lunch time or time to go home. Even without the sun he seemed to be able to judge the time of day with extraordinary accuracy.

I remember him telling many stories, especially the one about taking the quick cure for colds and flu. It was simply to take two drops of a well known cattle medicine on a sugar knob. The medicine itself was a kind of cure all which could be used as a drench when mixed with a pint of old ale or as lubrication if mixed with linseed oil but it was advisable to wash your hands thoroughly if used in this form. As Bristle judged his cold was particularly heavy, he said he upped the dose to three drops and within half an hour he was so hot he didn't know whether to send for the doctor or the fire brigade.

As with cricket, the Uffington teams were transported to away fixtures as far as East Hendred, Bishopstone and the outskirts of Swindon by Bob Freeman's village bus which he started operating in Uffington when he was demobbed following the First World War. His first conveyance was a rather flimsy looking grey coloured Ford which carried about a dozen people, seated on two bench type seats one on each side of the vehicle which meant that everyone slid to the front in the event of a sudden stop or to the back when going up hill. Another Ford replaced the original. It was a colourful yellow and blue, definitely more robust than its predecessor and as the seats were hinged to the sides it provided more floor space to enable churns of milk to be transported direct from local farms to the Express Dairy depot in Faringdon where it was processed and distributed, instead of being sent by rail to London . With the addition of a few panels of three ply to protect the glass windows it was also used to transport calves to Swindon market on Mondays at two shillings per head. Removal of the panels, a quick hose down, followed by re-adjustment to seating accommodation and it was ready and available for Choir outings, local shopping expeditions, Swindon football matches at the County Ground and similar excursions. Bob Freeman was a great supporter of "The Town" as well as being a useful player himself, I recall him driving the Uffington team to an away fixture and turning out on more than one occasion for them if they were short.

Following a few seasons of football with Uffington I spent two or three seasons playing hockey with the Shrivenham Club. They ran a ladies and men's team with an occasional mixed XI usually consisting of six men and five gals. Good fun it was too. We played on the Shrivenham Recreation field at the rear of the Shrivenham Memorial Hall. The Club was well known for the Six-a-Side tournament it staged annually at the end of the season which attracted teams from a wide area. The men's section often being won by "The Moonrakers" composed very much of Wiltshire County talent, while the Headington Ladies usually dominated the battles of the fairer sex.

About 1933 a successful attempt was made to form a Rugby Football Club at Faringdon. Its formation was due mainly to the hard work of Bernard Cook, the elder son of Billie Cook, the proprietor of the chemist's shop in Faringdon market place, now trading as Dow Chemists. Bernard was supported in his endeavours by Jimmie James' who had just been transferred from Oxford by Lloyds Bank to become head cashier, at their Faringdon Branch. Although admitting to be past his sell by date as a player, having been bred and born in Oxford and a passionate follower of the Dark Blues in cricket and rugby, he was willing and able to do anything he could to help. In fact he became the newly formed club's first Hon.Sec. and later also acted in the same capacity for the Faringdon & District Cricket Club for quite a few seasons. By kind permission of a Mr Cox, who farmed at Church Path Farm, we played rent free in a field of his which commanded a magnificent unhindered view northwards across the Thames valley to the Cotswolds. On most days conditions were ideal but if the wind happened to be in the North-East it was about the coldest spot in Berkshire. The Club headquarters were, initially, at "The Wheatsheaf" in London Street but moved later to "The Crown" which had just been purchased by Bunny & Queenie Taylor from Walter Ballard who had been "mine Host" for 43 years, and whose younger daughter Mary married a cousin of mine Norman Wheeler. They have a daughter Janie, two sons John and Jim and farm at Spanswick, Letcombe Bassett. At the time when their son Jim was due to arrive in this world, arrangements had been made for Mary to be confined in the maternity wing of Wantage Cottage Hospital. Unavoidably, the time for departure from Spanswick had to be in the early hours of the morning of March 6th in the year of our Lord, 1947. To make matters worse there had been an exceptionally heavy fall of snow during the night which made the journey by car almost impossible. However, a start was made on the 2½ mile journey at about 5.a.m. Progress was slow and by 6.a.m. had only reached the halfway point on the other side of Letcombe Regis when Norman enlisted the help of one of Sir Roger Chance's farm workers who was on his way to work. He was able to provide a trusty old carthorse suitably equipped with trace harness which towed the car through the drifts of snow whipped up by the wind on the open road to the south of Kirklands, to arrive at the hospital, just in time at around 7.00.a.m after what could understandably be described as the journey of a lifetime. Which was followed almost immediately by a joyful announcement that mother and baby were doing fine.

Unfortunately the Faringdon R.F.C. was unable to survive more than four seasons due to the difficulty of finding enough players in the locality and the occasional embarrassment of turning up one or two short. Had the Royal Military College of Science arrived on the scene at Shrivenham a year or two earlier, I'm sure we could have kept going but alas. Nevertheless we had a great deal of fun playing against teams such as Oxford Nomads, The Exiles, Newbury, Witney Reading, Stow on the Wold, Marlborough and Swindon whose old Black Horse ground is now underneath the M4 close to Wroughton. We were often supported by a few farm pupils while they were gaining farming experience under the much sought after guidance of the well known farmers Messrs Hobbs & Davis of Kelmscot. One pupil told us that while five of them were being taken round the farm for the purpose of explaining the advantages of rotational cropping, Robert Hobbs posed the question to one of them, as to what he considered would be the most rewarding crop to follow at this particular point. "Pupils Sir", was the quick reply. The local Vet — Townley Filgate was a real "terrier" at scrum-half. Unfortunately he was unable to play in away games in case he was needed to sort out a difficult calving or the like. Another stalwart was Bill Baylis who farmed at Hatford; he led the pack and had the distinction of playing in every fixture the club ever played. Which must have been well over fifty consecutive appearances.

It was while playing for Faringdon that my passion for watching rugby football developed and apart from the war years of 1939-45 I missed very few Internationals and Varsity matches at Twickenham. In the early years two or three of us would drive up by car, stopping off at "The Dumbell" on the outskirts of Maidenhead, a "watering hole" which every other rugger fan seemed to frequent. Fortified by a couple of 'halves' — not any more, as we were non-ticket holders, entering the ground through the turnstiles to view the match from the standing area as near to the players tunnel as we could get. Once there, a call of nature meant that such an advantageous view point would be lost beyond recall, as there was no way one could regain it by pushing back through a few thousand people who had built up on the terrace around us. To be close to the tunnel added tremendously to the atmosphere and enjoyment of the game. Close-ups of players and various dignitaries Royal and famous to whom the teams lined up opposite one another were being presented before the match began, coupled with the playing of the National Anthem, when the only movement in the ground came from the flags fluttering at the mast-heads of the two countries competing, always seemed to reach the very depth of my emotions. What a feast of names and memories it provided. W.W. Wakefield "Wakers" as he was called H.G. Owen-Smith "Tuppy", Prince Obolenksy "Obo" and of course many more. I was there when Obo scored his memorable but somewhat unorthodox try in the victory over the "All Blacks" in 1930 before a crowd of 63,000. Another great try that afternoon was scored by H. Sever of Sale. I recall him getting the ball almost in his own half within two or three yards of the touch line in front of the west stand and with little room in which to manoeuvre forced his way straight down the touch line leaving a string of defensive tackles in his wake to score in the corner in front of the North Stand.

In 1936 I was fortunate to be invited to go to Murrayfield to see England play Scotland, by a friend of Bernard Cook's, a Scotsman, named Ken Kerr. He was a partner in the Faringdon dentist practice belonging to Douglas Leahy and engaged to Bernard's sister Muriel (she always told me I was the first boy she ever kissed!) Ken intended motoring to Edinburgh to bring his mother down to Faringdon for a few days so arranged the trip to coincide with the week end of the match. He had swapped cars with Douglas Leahy for the purpose, as his M.G. Magnette was hardly an elderly lady's mode of travel and, of course, too small for the four of us on the return journey. So after a couple of beers in The Crown we were waved away on our 400 mile overnight journey in Douglas' imposing looking Sunbeam Talbot saloon at 9.00.p.m. complete with a few sandwiches, a dozen bottles of beer and no tickets — Ken said there would be plenty of standing room on the huge bank opposite the main stand. It wasn't long before we received signals as to what an eventful journey it was to be. After crossing the Thames at Radcot, having travelled about three miles, the lights began to flicker almost to the point of extinction. A call two miles on at the Ram-Jam Garage at Clanfield, the proprietor of which was well known to Bernard, a considerable amount of investigation failed to reveal any evidence as to why this should happen but somehow power returned to lights and starter motor although the garage owner freely admitted he knew not why. Our offer to reward him was turned down, saying he couldn't honestly charge us as he didn't know what he had done. So with our thanks and his good wishes we journeyed on, only to be plagued by the same symptoms while passing through Chipping Norton at about 11.00.p.m. We were amazed to find a garage with lights on and petrol for sale on the left hand side just after passing through the market place. Suffice to say our visit was a repeat of the previous one at Clanfield and despite extracting a mechanic from the pub opposite to assist; current seemed to be restored and we left again with good wishes for the remaining 375 mile journey and the added comment "Rather you than me".

All seemed very much better and we made good progress until having to stop at a level crossing near Leicester. While waiting for the gates to open, the engine stopped the lights went out and that seemed to be that. However, a bottle of beer and 20 minutes later a turn on the exceptionally long starting handle which we had recently discovered could only be inserted after a piece of metal cowling with the number plate attached was removed and up she struck once more; lights and all. Our next stop was north of Ollerton on the Nottingham/Doncaster road at about half past one in the morning by which time it was getting a bit misty and beginning to freeze a little. This hindrance appeared to be even more serious as we slowly petered out going along. Our efforts to attract the attention of the odd passer-by met with no success at all, due to the fact that there had been an increasing number of hold-ups reported recently in the national press and motorists were wary of being flagged down. Eventually, however, a man in a Morris Eight passed by us and reacted to our waving by stopping and reversing towards us. Satisfying himself before he got out of his car that we were in genuine trouble he produced a torch and started investigating under the bonnet, at which point Bernard whispered to me that "his shoes looked a bit oily, maybe he had something to do with a garage". How right he was and how lucky we were. He diagnosed dynamo trouble and a couple of faulty fuses. The latter he corrected by wrapping silver paper from a Players cigarette packet round them as a temporary measure and used part of the packet to pack the dynamo brushes tighter against the commutator. If I remember correctly, the starting handle fitted directly into the dynamo, so a few turns provided enough current to start the engine and after a few minutes on charge, the lights were operative as well. Unbelievably our troubles were still not over, as walking round the front of the car to get in the passenger side, Bernard noticed a trickle of water coming from underneath the car. Frantic investigation traced its origin to a hole in the bottom of the radiator cone, obviously caused by one of us when trying to line-up the handle into its socket in the dark. Once again our good Samaritan came to our aid, firstly by confirming our suspicion, not only that he was a motor mechanic but owned a small garage in Worksop about 9 miles away but off-route for us. He suggested however that we detour; he could then tip a couple of tins of Radweld into the radiator which would probably cure the leak and at the same time fit two new fuses. This was quickly agreed to, we pushed the car about half a mile further down the road, which was fortunately slightly downhill, to where he knew there was a bridge with a small stream passing under the road. Here we formed a chain to fill the radiator with water dipped from the stream in beer bottles, topped them up for replenishment when needed which turned out to be a bottle every two miles, until we arrived at his home and adjoining garage. As he ushered us into his sitting-room it was obvious that his wife who had been sitting-up for him expected him home very much earlier, was much displeased.

However, his recital of our misfortunes and intention to get to Murrayfield in time for the match backed by our expressions of thanks for his great kindness to us, soon mellowed her attitude and she quickly made us coffee which we gratefully consumed while her husband was doing his best to get the Talbot in fit enough state for the rest of the journey. His efforts completed and paid for in generous terms, including a topping up of the petrol tank we drove out of his garage into one of the thickest fogs I have ever experienced which enveloped us in varying degrees all the way to Princes Street, Edinburgh where we arrived at 11.30.a.m. and the fog was still so thick we couldn't even see the Castle from the street below. Our fears that the match might be 'off' were dispelled about an hour later when miraculously the sun broke through burning up the mist quite quickly making conditions ideal and the prospects of a great game even more exciting.

Thanks to our garage friend the car behaved itself and apart from a Police Patrol man walking into the cafe at Borough Bridge , where we had stopped for breakfast, to inquire who the Sunbeam Talbot with no front number plate belonged to, due to us having chucked it in the boot on leaving the level crossing at Leicester; apart from the fog the last two hundred miles or so were free of incident. The match itself was a good one and it was great to watch England under the captaincy of Tuppy Owen-Smith lay the Murrayfield bogey after a run of so many defeats on Scottish soil. On the way back to Ken's mother there was enthusiastic talk of joining celebrations in the city during the evening but his mother's high tea changed all that; we hit the sheets at 9.p.m. woke up at 9.a.m. and were back home in Faringdon during early evening after a very pleasant return journey, which included a stop for lunch in Doncaster. All in all it was a memorable experience, and has been talked of many times since.

It was nice too that we had a note from the garage owner's wife, in response to sending her a box of King George V chocolates on our return home.

Further sport was also available in the form of fox hunting with the Old Berks. In those days most farmers had some kind of nag on which they could follow hounds for an hour to two, once or twice a week. A few also bred and kept better class horses to hunt and "sell on" or perhaps run in a few local point-to-points within hacking distance. My enjoyment of fox-hunting varied for one or two reasons, the first being that I was becoming increasingly conscious of the financial state of farming and felt I should be at home trying to keep the wolf from the door. Secondly, father loved horses, apart from breeding a few which he usually sold to the Army Remount Depot at Arborfield as unbroken 3 year olds. He was also very fond of buying and selling — although there were times when I used to think he was not particularly good at it. This meant he was quite often stuck with what was known in the trade as a "bad un" and there were occasions out hunting while trying to cope with such an animal that I came to the conclusion that there were other ways of enjoying oneself. Nevertheless, I still admit there was nothing better in the world than being on a "good'un" — the only trouble was, it soon belonged to someone else. As for point-to-point riding, I never had the courage to think about it, let alone do it. Perhaps subconsciously I was influenced in this respect by my mother's luke warm enthusiasm for matters equine. She rarely accompanied father to point-to-points when he was riding, for the fear, I'm sure, of possibly seeing him fall and being injured, although like many farmer's wives she accepted the hunting scene as an integral and almost necessary part of country life and probably suffered more than a little in silence.

My last day hunting in the saddle followed a lawn meet of the Old Berks at Major Nickisson's home, Hinton Manor, in the mid 1930's. It was on a mare named Alice, provided for me by my uncle Jack Wheeler of Grounds Farm, Uffington. We had a good day with two or three useful runs finishing up at Ringdale near Fernham just before dark.

Mention of Uncle Jack prompts me to record an extract from an article which appeared in a monthly journal entitled "Caravan". It was written by E.E. Kirby in 1960 and describes a caravaning holiday spent in a field at Grounds Farm and a lasting friendship which developed with Uncle Jack during the many holidays spent exploring the Vale of the White Horse. The author describes Uncle Jack as being a farmer of the old school saying amongst other things that he disliked social pretensions and often deliberately set out to shock when he met them. The author went on to say and I quote " that he remembered one occasion when they were having tea at the farm during a tennis party. Jack came in sweaty after hand milking, found no place laid for him at the table, and snorted forcibly. 'Ah! more pigs than titties, I see. More pigs than titties'. One would have to go to Shakespeare to match that metaphor. Didn't Falstaff moving ponderously in front of his tiny page, declare 'I do here walk before thee like a sow that hath overwhelmed all her litter but one!'

It so happened that most of my comparatively short period as a mounted follower was in the 1920's during the mastership of Dr Paget Stevenson and his famous kennel huntsman Fred Holland, who was very ably assisted by his whipper-in Jim Goddard whose holla never failed to make the hairs on the back of my neck stick out and send an exciting shiver down my spine. To the best of my memory the size of the average Field would be around seventy followers, almost half of whom would be farmers or members of their families, which meant that hounds were welcome almost everywhere and no-go areas could be counted on the fingers of one hand.

Fred Holland was considered by most people who know anything about hunting to be a "natural" and stood out like a piece of Chippendale in a roomful of G.Plan. I remember being out on an old cob with father on a day that had not provided very much. After a short run, a fox found in the old canal near Uffington, had been marked by hounds into a stone drockway close to Common Farm and was proving difficult to bolt. It was about 3pm and Fred being anxious to provide some fun before the end of the day had asked the Master if they could move on, but without success. He had also heard father say to me that we had better get off home to milking, so he asked us to "Holler like Hell" when we reached Baulking railway bridge about half a mile away. This we did — Fred, shouted immediately "Somebody has seen one master, can we go?"

which was this time answered in the affirmative with the result that Fred and his hounds were almost "hunting" when he crossed the bridge to shout "Which way has he gone Cyril?" "Down the green past the Church," replied father. Whereupon Fred galloped off towards Baulking Copse less than quarter of a mile away from whence a surprised "Charlie", was soon afoot to provide a lively half hours hunt before getting to ground at Rogues Pits near Shellingford to be left for another day.

The G.W.R. London to Bristol main railway line could be dangerous if a fox crossed over it, but most steam trains didn't travel at express speed which was reckoned to be 60 m.p.h. so quite often a driver was able to slow down or bring his train to a halt should he see hounds on the line ahead of him. Such demonstration of goodwill was always rewarded or acknowledged at the appropriate level. The Faringdon branch line from Uffington Junction posed very few problems although it ran through four or five miles of good hunting country and dissected a well known draw; Coles Pits. The Faringdon Johnnie, as the train was nick-named by the locals, usually consisted of two non-corridor passenger coaches capable of carrying about 50 passengers each, two or three goods vans, an open truck or two for coal etc., a milk van, ventilated by 3 inch spaces between the wooden boards forming the sides of the van, to keep the milk cool in transit. This had the reverse effect in hot weather of allowing the sun direct contact with the outside of the churns. Hitched on to the rear of all this was the guards van. It was rather like a small travelling office equipped with a stove and small chimney pipe for winter warmth. This area had access through a rear facing door to what I can only describe as a covered veranda from which the guard could lean out either side to see if all was well and wave his red or green flag, if necessary, to communicate with the driver and his fireman. The turn-round at the end of each journey necessitated a fair amount of shunting at both ends of the track in order to get the engine in front of the guards van at the rear. Each single journey took about 15 minutes which indicated quite a leisurely speed.

It was said that on "goods only" journeys the driver sometimes halted the train in the Coles Pits cutting which due to its somewhat sandy soil was well stocked with rabbits, and was therefore an ideal area for the setting of a few snares, the produce from which being picked up on the return journey.

There was also an occasion when a passenger from Faringdon pulled the communication cord in the vicinity of Fernham Bridge, saying to the guard as he climbed down on to the track that he had come far enough, as he was visiting his Auntie at Fernham. Another incident concerned the Reverend Cowell, an erstwhile vicar of Faringdon, who with one or two others was left sitting in a coach coupled only to the guards van while the rest of the train proceeded to Uffington Station. As he was due to attend an important meeting at Lambeth Palace, the engine was uncoupled on arrival at Uffington and raced back to get him. The main line connection for Paddington was held up for him and the day was saved.

My mother was also the recipient of V.I.P. treatment when she decided to visit Faringdon for Saturday evening shopping ahead of father who was to follow in the car later. Most shops stayed open until 8 p.m. to accommodate customers from surrounding villages. Having walked half a mile to the station on a very foggy November evening to catch the 5.15.p.m. the station-master Mr Williams informed her that all main line trains were running at least half an hour late and the situation could worsen if the fog thickened. Anxious to oblige, he said as the engine was more or less at the ready, rather than stand idle there was plenty of time to run her into Faringdon and return, so he sold her a sixpenny single ticket, escorted her to the carriage,, collected the ticket back again, gave instructions to the three man crew and off they went. I recall mother saying how embarrassed she felt alighting on her own at the other end and hearing someone say "wonder why they haven't turned up — hope they are alright".

I have always been a supporter of Point-to-Point racing. My

interest in the sport began when my grandfather was Hon. Sec. of the Old Berks fixtures in the 1920's, which were very much part of the life of the locality during the spring of the year. Other fixtures always attended were those of adjacent Hunts such as The Beaufort, Craven, Vine, V.W.H. Cricklade and V.W.H. Bathhurst. The days for these events, were usually chosen to coincide with the nearest town's early closing day and was an occasion when local farmers and tradesmen were entertained very hospitably at the Master's tent and at farm wagons occupying vantage points, hosted by well known members and subscribers to add individual thanks to those of the Master for the privilege of riding over farmers land.

The "Old Berks" fixture was run almost entirely by voluntary subscription, there were very few cars, so no revenue from car parking, race cards sold for 6d. (2½p). Grampy's account book for 1926 records a revenue from this source of £30. On the debit side 9 weeks wages for course building totalled £19.13.8. Police presence for the day cost £8.1.8, tents £4.10.0. The net result of all the work and effort was a profit of around £200 annually which was paid into the Hunt Wire Fund for the purpose of improving the country and making it safer to ride over.

Memories of being a spectator included seeing H.R.H. Edward, Prince of Wales finish second on a mare named Little Christy at the V.W.H. meeting at Blunsdon, as well as seeing Bruce Hobbs win a race over the Step Farm course at Faringdon when he was sixteen years old, not long before he won the Grand National at Aintree on that good little horse Battleship, trained by his father. On the same afternoon Mrs Aitken, a lady past middle-age, from Southmoor, finished third in the Members race, riding side-saddle. She would have been second, but was baulked at the last fence, by farmer Billy Cooper's horse refusing, bringing a spate of comment not particularly in keeping with a lady attired in riding habit, silk topper and veil.

Mrs Aitken's dignity also took a bit of a knock, when

Viscount Barrington hosted a meet of hounds at Becket House, Shrivenham. As usual her groom had hacked her horse on to the meet in order that her husband, a tall aristocratic gentleman of few words and sadly a little deaf, could drive her along in his Model T Ford Saloon later. For the purpose Mr Aitken who had positioned the car close to the front door of their house at Southmoor, opened the rear door ready for his wife to get in as, being attired in a riding habit, she preferred the extra room in the back. It so happened it was rather a windy day and the sound of the rear door shutting, led him to believe as he sat in the driver's seat, with the engine running, that his wife was safely aboard. So without a glance over his shoulder he drove the 8 or 9 miles to Shrivenham unaware that he was on his own. I am not aware that there is any record of what was said when he returned to Southmoor to pick his dear wife up!

There is a story attributed to the Old Berks Point-to-Point when it was held at Baulking prior to the First World War. The course required the famous hunting hazard, Rosey Brook, to be jumped twice. With the water dammed up and no vestige of a fence on either take-off or landing side, it was a tricky obstacle to negotiate. Story has it that an unfortunate horse and rider failed to make it on this occasion, going in over head and ears. After considerable thrashing about in the water, a lone spectator standing nearby rushed to the rider's rescue by pulling him up the bank as far as he could and started doing his best to revive him by pumping up and down on his chest. After some minutes, with water still coming out of his mouth, nose and ears he was aware of someone galloping towards him on horseback to see what had happened. Being almost exhausted the spectator, shouted "Come quickly, I need help — do you know anything about artificial respiration?" to which the chap on the horse replied "No — but I do know a little about elementary hydraulics and if you don't lift that poor so and so's backside clear of the water, you'll have the brook dry before long!"

Social life during winter months consisted mainly of a few

supper parties involving enough guests to make up a couple of tables for "Ha-penny" nap afterwards, the ideal being five players at a table or six if the dealer sits out and pays or receives when it is his/her turn to deal. A hand only takes about 5/6 minutes to play and doesn't require a great deal of deep thought, so there was always ample opportunity to chatter while the game was in progress. As with every card game the extent of the gamble is relevant to the stakes agreed upon before the first hand is dealt. A single "nap" hand, winning 6d. from each player was usual, so if you were bold enough to call 6 times nap with six people playing and won the hand it was worth fifteen shillings; if lost it cost the caller seven shillings and sixpence as losses were paid at half the winning rate. At the end of the evening at this stake level it was seldom if anyone won or lost much more than a £1 and a good time, accompanied by a drink or two had been had by all. Although I do remember one evening at Church Farm when three guests, two of whom were Scotsmen and of course drinking whiskey, acted like the real gentlemen they were by refraining from comment and even came back for more when father, who as usual was drinking gin, gave them sherry in a whiskey glass, inviting them at the same time to say how much water they wished to have with it! All because mother had inadvertently changed the positions of the sherry and whiskey bottles in the decanter when she was topping them up earlier in the day.

There were also a few local concerts to go to, some of which were held in what were known as Village Huts — rather than Halls. Most of them were purchased from the Army Surplus Depot at Didcot just after the First World War. They were a bit crude, the interiors being made of wood sections and exteriors of corrugated iron sheets, but with a little ingenuity they provided adequate accommodation for village socials, dances whist drives and the like. Baulking raised sufficient money to buy one but it turned out to be a bit of a disaster as, due to strong opposition from a minority of common holders, it could not be sited on any part of the village green, so it was erected by permission on a piece of land adjoining the main Great Western Railway line, having been purchased by the G.W.R company when the line was laid and opened in May, 1841, and is still within their visible boundary mark. Despite a great deal of effort and good intention it was never a success, due mainly to the fact that it stood on the worst possible site in the village. Being so close to the railway line and within a couple of yards of the railway bridge, the noise from passing trains especially long goods passing under the bridge, made it impossible for those inside to carry on a conversation for five minutes at a time and, if the door happened to be opened and the wind was in the wrong direction, the smoke from the steam engines could be a bit tiresome as well. On one occasion the Faringdon Orchestral and Choral Society had to abandon its effort to entertain a full house following frequent interruptions due to noise. This made it impossible for the conductor Mr Bernard Haines, a Faringdon Solicitor, to co-ordinate what was happening as, when a train had passed, everyone under his baton seemed to be a bar in front or behind, the final straw coming when an express had to wait under the bridge for quite a while for a signal to change to all clear. Our Vicar at that time, the Rev. G.H.C.Bartley played second violin in the Faringdon Orchestra and I remember the pianist Billie Cook telling me of Mr Haines tapping his music stand to bring his orchestra to attention for the first piece on the programme at a concert in the Corn Exchange at Faringdon. Silence fell upon the audience as Mr Haines launched them into action with a flourish of his baton. After a very short while and a few anxious glances in the vicar's direction he ceased conducting, tapping the top of his music stand to bring the playing to a halt. He then approached Vicar Bartley to examine the piece of music he was playing, which resulted in it being quickly replaced by another from the portfolio in his music case by the side of his chair and, with a reassuring pat on the shoulder from the conductor, the Reverend Bartley made sure during the rest of the evening he was playing from the same score as everyone else.

Mr Bernard Haines was a solicitor of the old school, a kind and thoughtful man who followed in his fathers' footsteps in premises now occupied by the Cheltenham and Gloucester Building Society. A visit to him transported one's imagination to what it was probably like in Dickensian days. The large table in the middle of his upstairs office was stacked high with legal documents tied up in bunches with pink ribbon and it always amazed me how he knew instinctively which bundle to pick up in order to deal with the particular matter on hand. If a signature was required, it was usually difficult to find space on the table for the document and was often carried out on a thick bed of correspondence in the process of being dealt with. He employed a clerk named Jack Sollis who sat or stood at a very high desk and worked for Mr Haines all his life, barring the first World War. He played cricket for the Faringdon branch of the British Legion, and was a useful bowler and I recall playing against him at Uffington a few times. Mr Haines himself was a keen sportsman in his younger days, being a good cricket and tennis player and, as President of the Faringdon and District Cricket Club in his later years, he rarely missed walking up to the field at the top of Southampton Street to sit in front of the old thatched pavilion to watch the game and chat about the chances of England winning the Ashes and similar topics.

Sadly he died tragically having fallen into a ditch by the side of the road in the dark about a mile out of Faringdon where he was discovered the day after. Having failed to get off the Swindon bus near his house in Coxwell Road, he was apparently walking back home from the next stop when the tragedy occurred.

His father Alan, was involved in an amusing piece of litigation, which happened round about 1908 and I remember my grandfather telling the story with relish more than once, as it concerned a close friend of his, Edward Heavens, who owned the butchers shop close to Portwell in Faringdon Market Place, which still trades under the name of Pat Thomas Ltd. Most butchers shops in those days were open fronted, displaying meat on marble slabs and it so happened that two very boisterous unattended half grown hound pups managed to grab a prime leg of lamb from Edward's display. Before he was able to stop them they galloped across the market place, finally disappearing up Southampton Street pursued by him but without the slightest chance of being caught. Eventually losing sight of them, he returned to the market place in a very angry mood. Determined to exact some sort of retribution he went straight to Alan Haines office where he was persuaded to calm down a little and explain what had happened. Assuming Edward needed his help in the matter, he inquired the value of the stolen property which was estimated at two shillings and sixpence. Alan said the next step was to find out to whom the puppies belonged. Edward replied that technically he supposed they were the property of the Old Berks Hunt but no doubt they were being cared for by a puppy walker. "Have you any idea as to who he might be?" inquired Alan, "Yes, you" said Edward "Ah well in that case replied Alan, my standing charge for advice is three shillings and sixpence so if you give me a shilling we can consider the matter settled. Incidentally, Edward and his son George, who later took over the business could often be seen sharpening their knives on the edge of the stone water trough known as Portwell which formed part of the drinking place erected by the Pye Family from Faringdon House for the purpose of providing water for householders without a supply of their own as well as to thirsty horses and cattle, especially on Market days.

 Cinema entertainment in Faringdon was available up to about 1930 in the Corn Exchange and I recall being taken as a small boy to see "The Kid" a silent film starring Charlie Chaplin and Jackie Coogan. I forget the story, probably because I wasn't quick enough to read what they were saying on the screen, but I can well remember the constant clicking of the projector while the film was in motion and what appeared to be continuous lines running down the screen.

The charges were sixpence up front, a shilling at the back and one and sixpence on raised platforms either side of the projector box. These two areas were known as Cosy Corners and were frequented by those interested in a bit of courting. One of the last proprietors was a Mr. Barnard who accompanied the films with music from a piano and a small harmonium, which he cleverly played by sitting between and using one hand on each to emphasise the mood of what was happening on the screen.

With the cinema fast becoming a popular form of entertainment and the possibility of talking pictures becoming a reality, a purpose built cinema was created on a site opposite the Corn Exchange next to "The Volunteer". It was completed in the mid 1930's and opened by a well known silent film actor Stuart Rome whose sister married George Heavens, the Faringdon butcher. Sadly with the coming of television and the decline in small town cinema going, despite a period as a Bingo Hall it was pulled down to provide a site for housing accommodation..

Before it closed as a cinema, a mini drama occurred one evening concerning Tommy McBain, the landlord of the "Junction" public house at Uffington Station. His son Joe announced he was going to Faringdon on his motorbike. A younger son David took the advantage of a lift on the pillion for the purpose of going to the cinema. As Joe was going elsewhere after Faringdon it was arranged that David should catch the last bus to Uffington where his father would meet him. When the bus arrived David was not aboard. His father thinking he had probably altered his plans and would return with his brother Joe, panicked when Joe also arrived home without him. It was getting late so father went straight to Faringdon Police Station from where he was accompanied by the Officer on Duty, Fred Butler, to knock on the door of the cinema owner, a Mr Pawley, who was adamant that there was no-one left in the cinema when he locked up. Nevertheless, he was persuaded to go with them to be absolutely sure. On opening up and switching on the lights, close to midnight, there was great relief all round when the curled up body of 12 year old David, still asleep was found in the "nine-pennies" — three rows from the front.

In the early 1930's The "Regent" Cinema was built in Regent Circus, Swindon. It was here that I saw my first "talkie" "Bulldog Drummond" starring Ronald Colman and I think Claudette Culbert. The "Savoy" at the top of Regent Street near the Riflemans Hotel followed shortly afterwards and with the Empire Theatre provided a variety of entertainment for the purpose of a family outing or taking a young lady out for the evening.

The Empire was my favourite as it presented a mixed programme consisting of Edgar Wallace thrillers, straight plays and farces and a good selection of variety shows, incorporating all kinds of turns, many by top of the bill artists. I remember being daft enough to answer the call on two occasions for a member of the audience to go on stage. The first was when Davant the surviving member of Masklyn and Davant, the famous illusionists was appearing. He called for two people from the stalls to inspect a strong wooden framed chest with glass panels in the lid and sides. It measured about 3ft x 2ft x 2ft, the lid being secured by two locks and keys. In his patter he said he had performed this trick with this very same chest in front of Queen Victoria on the lawn at Windsor Castle. We had a good look at the chest inside and out and checked the security of the lock. At this point he brought on an attractive scantily glad young lady who just managed to curl herself up enough to fit into the chest, the lid was locked by both of us and we put the keys in our pockets. Davant then gave us some three inch wide webbing with metal rings on the end and some short lengths of cord with which to tie the rings together and invited us to parcel the chest up in any way we pleased to make sure the young lady could not get out. Satisfying ourselves we had done our best, a canvas cover about the size of an old fashioned bathing tent was placed over the young lady who smiled at us from within the trussed up chest as the curtain was drawn across the front, we were requested to stand a yard or so back, one on each side. Following a wave of his hand, a small explosion and a puff of smoke, we were invited by Davant to pull back the curtain and there sitting on the lid of the chest which was still locked and parcelled up just as we had left it, was the charming young lady waving to the audience. Davant thanked us for our assistance and we returned to our seats without a clue as to how he had done it.

Maybe the experience wetted my appetite, as not long after I answered a similar call — or so I thought — from a performer whose name I cannot remember. His act consisted of various Juggling and Balancing feats. To round it off he welcomed me on to the stage saying all he wanted me to do was stand absolutely still and, providing I did so, there was no need to worry as up to now he hadn't maimed anyone for life, or words to that effect. So it was with a fair degree of apprehension that I carried out his instructions, which failed to reveal the object of this particular part of his performance until the very last moment. He stood me on one side of the stage, balanced a lighted candle on top of my head, clipped a cork with two cigarette cards slotted into it in the shape of a mini aeroplane propeller, on to the end of my nose and then asked me to stretch both arms out in front and grip a cigarette card between the third and fourth finger of each hand. Apart from feeling a right Charlie I still hadn't a clue what he was going to do, until he took a long piece of cord with a billiard ball attached to each end of it and after dropping them on to the stage floor a time or two to prove that they didn't bounce and were perfectly genuine he started swinging them round about his head in opposite directions in lasso fashion. Again warning me to keep absolutely still he positioned himself about 8 feet away from me and proceeded to lengthen the cords gradually through his hands until the balls knocked the cigarette cards from my fingers. Further applause indicated that he had also extinguished the flame from the candle in similar fashion. Resorting to the use of one ball only the same process removed the two cards from the cork on my nose and before I could relax he said, "stand still, I'll take the cork as well," which to everyone's amusement and my relief landed half way into the stalls. On leaving the stage he invited me and any friends with me to have a drink with him in the Stalls bar during the interval. So four of us accepted the invitation which was a short but interesting experience as he introduced us to others appearing on the bill. Those I remember were Billy Russell billed as the working man's comedian, who told us his missus was a big woman, and he loved every acre of her, also two Canadian crooners named Al and Bob Harvey. Crooning was a style of singing that had become very popular at that time due to the arrival of the young and talented Bing Crosby on the entertainment scene.

During winter months there were a number of Balls and Dances in the area which one could attend. They were mostly held in The Shrivenham Memorial Hall or the Faringdon Corn Exchange and organised by local sports clubs mainly for the purpose of raising funds as well as providing a social side to their activities. Tickets for Hockey, Rugger and Cricket Dances were usually five shillings which included a buffet supper. The Faringdon Bachelor's Ball and the O.B.H. Wire Ball being ten shillings and by invitation only. The Hunt Ball was an event apart and attended by Hunt Subscribers only. Three of the bands I remember were Jack Viners of Witney and The Blue Star Players from Oxford and George Whitfields's of Wantage.

There were a large number of Hotels/Pubs per capita of Faringdon's population. The Bell, under the landlordship of Bert Franklin, seemed to cater for the business men of the town, while The Crown under mine host Walter Ballard was frequented mainly by the local farming fraternity, some of whom were great characters and indulged in a fair bit of leg-pulling. I recall Stanley Adams, a son of the legendary George Adams who farmed a very large area of land around Faringdon at the turn of the century, saying to me in The Crown early one Saturday evening that old Albert and Howard, two fellow farmers, will sure to be in later. "We'll have a bit of fun, and when they have settled down on their usual bar stools together, you drop this large white silk handkerchief," which he pushed into my jacket pocket, "on the floor between their stools and leave the rest to me". Sure enough in they came and the bar having filled up I sidled up to them had a few words and, did as I was told. After I had moved away Stanley said loudly enough for most people in the bar to hear: "One of you two have dropped your handkerchief," and bent down to pick it up for them, making sure as he did so everyone could see it was in fact a scanty pair of French knickers. I must admit I was as surprised as they were but not half so embarrassed. It was accepted that the perpetrator was wide open to some kind of retaliation, as was the case when an advertisement appeared in the local press offering budgerigar eggs for sale and the subscriber to the telephone number quoted turned up in The Crown a few evenings later, saying he couldn't understand why people kept ringing him up for budgerigar eggs!

It was all taken in good part and provided something to talk about. Countryside humour was very much self made and off the cuff in those days, possibly resulting from the fact that much of the work on farms was performed by a number of people working together in a group which provided ample opportunity to chatter such as when hoeing, muck spreading or sitting on stools doing the hand milking.

The usual topics were booze, women or the boss, but occasionally bits of sensational news promoted considerable discussion, such as when Sir Henry Segrave broke the world land speed record, travelling at something like 187 miles per hour which prompted a remark that "He wouldn't half have come a gutzer round by Uffington Church" — a piece of road notorious for its two S bends! On another occasion, having been to a dance and only had an hour and a half between the sheets, I was struggling to keep awake during early morning milking and doing my best to keep up with the others. My efforts had not gone unnoticed by Jim Loveridge, a strong gaunt man with a permanent twinkle in his eye. After milking about four or five cows each, he was sitting at the cow in front of me and must have looked over his shoulder to see that the warmth from the flank of the cow I was milking had momentarily lulled me into a state of inactivity and my hands had dropped on to the top of the bucket between my legs. Waiting patiently until he saw them starting to come to life again, using two forefingers he bent my cows' furthest front teat upwards pressing it against her udder. After watching my half-awake efforts to locate the "missing" teat, Jim said with a chuckle "Can't ee find un then, gaffer?"

One morning while I was littering up some calves, Jim came to me with a broad grin on his face saying, "Come quickly with me to the barn, I want to show you something." I knew father was in there rolling some oats, so I asked him if he was alright "Yes he's alright" said Jim "Come and see". When we got there father was lying on his back in a narrow walk way left between two stacks of stored grain in four bushel sacks, stacked two high. He was unable to get up, as he was pinned down with a 1½cwt sack of oats on top of him which he had taken in his arms off the top row of one of the stacks. As he was struggling backwards he caught his heel in another sack, causing him to fall backwards, still clutching the sack, which was the same width as the walkway. Jim said father had told him he was not hurt in anyway otherwise of course he would have got him out straight away so he thought it was too good a sight to keep to himself. I'm not sure that is what father thought, but he was none the worse and I heard him tell the story of his temporary perplexing situation more than once.

Another job that didn't mix particularly well with the after effects of a late night was steering a horse drawn Gilbert corn drill. The drill was patented and made at Shippon near Abingdon by John Gilbert, the father of Roy Gilbert who married a sister-in-law of mine. It was a versatile machine capable of sowing corn, root, grass and clover seeds. It required two horses to pull it, three if the ground was heavy and a team of three men, one to drive or lead the horses, one to steer so that the drills were kept at a uniform width throughout, and one to walk behind to drop and lift the coulters at each end of the field and make sure the seed was running properly. Steering was reasonably accomplished in good tilth but if weather conditions had been unkind and dried the clay soil into a lumpy seed-bed, then walking beside the drill hanging on to the steerage arm became difficult enough without having to steer as well and, at the end of a days drilling of around ten acres, mind and body were ready for a rest. The other farm task which produced similar physical fatigue was loading hay on to a wagon from a hay loader, which was hitched on behind the wagon and picked the hay up from a swath formed by a horse drawn side-rake putting two or more swaths cut by the mower into one, depending on the size of the crop, and the judgement of the chap sitting on the seat of the side-rake, which was moulded from cast iron and required a liberal amount of hay between it and the drivers' back-side to make it anything like comfortable. Most horse-drawn farm implements were fitted with similar seats, i.e. mowers, reapers, binders, side-rakes, swath turners and horse-rakes etc. Cultivators, drags, harrows, rollers, skims, horse hoes and the like, being all wheel-less required the driver of the horses to walk behind. It was not uncommon on quiet sunny days in springtime to hear workers whistling or singing a few fields away while following their teams of horses, seeming to convey a kind of joy and pride in what they were doing. To say that it was difficult to display the same kind of contentment when dealing with hay coming at you incessantly from a hayloader would be an understatement. I recall our carter, Bill Mace, telling me to load the rear of the wagon first, so that the hay formed a wedge down which it was easier to get it to the front and to keep you above your work and helped to bind the hay together, to prevent it falling off the load, which spelt disaster. A slipped load meant that probably half a ton, or more had to be pitched back on the wagon again by hand. If there was any contentment attached to the job, it was lying flat on your back on the top of the last load of the day, gazing into the evening twilight while being transported slowly to the rickyard and carthorse stables conscious of the fact that when you had helped feed and water the horses, often by the light of a hurricane lamp, and turned them out into the home paddock for the night, that you had at least earned the crust of bread and cheese and bottle of beer that would be waiting on the supper table. On long days appreciation of willingness to keep at it while the sun was shining, took the form of a bottle of beer to everyone concerned when saying "Good Night".

I recall a couple of haymaking dramas. The first was in about 1925 when two cart horses careered into the stable yard attached to a pole and two wheels which was all that was left of a comparatively new mower, pursued by an out of breath Bill Mace, the carter, who explained that they had run away with him when the mower cut into a wasp's nest, he fell off, was stung a few times and the horses galloped home through three gateways leaving bits of broken mower behind them.

The other drama concerned myself and a character named Harry Eltham who worked a number of years for father as a groom and general farm worker. It was a warm and sultry afternoon in mid June following some morning sun during which I had turned the hay a couple of times and it was fit to pick up about 3 o'clock. Arriving in the field named Banny Hill with the wagon and hayloader, we both remarked that the sky looked a bit black over Challow Station way, an area from which Baulking folk always reckoned was a sure source for a first class thunderstorm developing. With a young horse in the shafts and a steadier type as trace horse and Harry as loader, I set off up the field, turned at the far end on the bare headland which emptied the loader and at that moment there was a blinding flash of lightening accompanied simultaneously by a deafening crack of thunder causing the horses to bolt immediately. With Harry shouting to me to hang on to them, I was helped somewhat in keeping up with them by the weight of the wagon, the hay on it and the attached loader. However, after about a hundred yards they were still in a lumbering gallop and as we skirted round the pond in the middle of the field my fear was that they would head for the gate on to the road a few yards from the railway bridge and the lot of us would be smashed to pieces. As I couldn't hear Harry anymore I assumed he had either baled out or fallen off. With the gate coming nearer, rightly or wrongly I let go of the halter on the head of the horse in the shafts and pulled as hard as I could on the off side rein of the trace horse. Thank God he responded veering to forard (as the old carters would say), taking the other horse round with him in a semi-circle. They finally came to a standstill, whereupon I looked back at the wagon to see Harry's face, albeit as white as a sheet, looking over the top of the hay, with his hands still clinging to the top rail of the front rave. His first words were "I don't know about you but I could do with a pint".

A quick look at the hayloader revealed that the damage was extensive, so haymaking was finished for the day and it was taken to W.Packer & Sons of Uffington for immediate repairs. The extraordinary thing was that it didn't rain and it was the one and only flash of lightening and clap of thunder that we had that day; although we learned later that Wantage six miles away had suffered torrential rain and flooding.

Harry Eltham was certainly fond of his beer and was a regular at "The Junction" at Uffington Station where he was often joined by the stationmaster, a Mr. Lee, who invested in a motorbike so that they could taste the brews a bit further afield. This rather prevented their wives' participation in a night out, so Harry bought a second hand B.S.A. of his own which cost him thirty shillings. After one or two trips down the village green he felt confident enough to take his wife for a ride astride the pillion, for the first time. In the presence of a few onlookers, of whom I was one, Harry revved up the machine in the road outside his cottage, dropped it into gear before his wife was ready with the result that he shot off over the railway bridge out of sight leaving his wife standing with her legs wide apart in the middle of the road. As he failed to return in a few minutes we began to fear the worst but eventually he turned up to say he was sorry to be so long but he had to go to Longcot (about 3½ miles) before he could turn round! He never did master the machine and after a short while went back to his old push-bike, which if he couldn't ride was at least something to hold on to.

At that time the milking sheds were open fronted and the milk from each cow was tipped into three large carrier buckets, with a capacity of about 3½ gallons each. When two of these were full it was Harry's job to carry them across the cobble-stoned cow yard to the dairy where they were raised to shoulder height and tipped into the milk receiver, equipped with a brass tap which controlled the flow of milk over the water cooler, before it entered the churn standing underneath. Harry's efforts, during early morning milking, to accomplish the task while suffering a hangover from the night before were often the cause of a fair amount of ribaldry but I never saw him spill a drop — of milk and he was never late. I liked Harry and I think he liked me.

Another very likeable character but of a very different ilk was Henry Ackrill who worked for father for a few years prior to retiring and carrying on doing a few odd jobs such as a bit of root hoeing when he felt like it or cutting the grass in the churchyard with a scythe, accompanying every cutting stroke with a kind of quiet whistle which he made by breathing out with his tongue pursed lightly between his lips. When I asked him why he did it, he said before horse drawn mowers became available all the grass for haymaking was cut by scythe and as well as being back-aching, it was a thirsty job as well, and the theory was that it helped to keep your mouth and lips from getting parched in hot weather. I have heard grooms and carters making the same noise when grooming horses with dandy brushes and curry combs. Grampy's groom, whom I remember as "Old Hawkes" told me that it helped to settle a horse into a state of calm and quiet which in time helped to build that bond of trust so desirable between horse and groom. To me Henry Ackrill was a representative of what I imagined farm workers living in Baulking were like in Victorian and Edwardian days. He was of average height, stocky, with a kind face adorned by a medium length, well trimmed greyish beard.; a steadfast, reliable and tidy man who took a pride in what he did. He wore a cap indoors as well as out and apart from being in bed the only time he removed it was for a short while every evening when he put it on the table beside him while he read a passage from the bible.

I never heard him swear, although occasionally he seemed to come very close to doing so with expressions such as "D....ally, now look what thee's done" or when supervising the building of a corn rick "B...eggar the boy, keep the ears pointing inards like I told thee". His working attire consisted of a sleeved jacket of thick dark grey cloth, cut rather like a waistcoat. His trousers were "yorked" as it was called, just under the knee, with a narrow leather strap, producing a knickerbocker effect. During the winter he wore gaiters to help cope with wet and muddy conditions and his leather boots would receive liberal dressings of dubbin to help keep his feet dry. Wellington boots did not come on the scene until after the First World War. Henry told me he and his wife brought up five children on twelve shillings a week and he remembered his father talking of the "hungry fortys" and how, when he assisted in killing a pig at the farm, he was grateful to be given the "trotters" to take home to boil for the tiny bit of meat on them and provide stock to make some soup for the children.

I remember one day when Henry was helping to move some heavy flag stones for the purpose of making a path, one of the stones fell on to his foot, causing the old chap considerable pain and making him feel a bit faint. So my father popped indoors and returned with a stiff tot of brandy saying "Drink this Henry, it will do you good" "Oh, I couldn't Master — I couldn't" said Henry "Well if you don't, I damn soon shall" replied father. Whereupon the old chap took hold of the glass for fear of not getting a further opportunity and almost downed it in one.

There were other similar families who had lived in the village for three or four generations such as the Gerrings and the Painters. As there were eight farms within the parish boundary, all with dairy herds, it was seldom necessary to seek employment elsewhere. They were all craftsmen who had learned their trade mainly by watching others, some preferring to work with horses rather than dairy cattle perhaps because carters had slightly more time between the sheets than cowmen — I don't know.

I also remember a jolly lady of ample proportions named Granny Iles who helped mother with the washing when we first came to Baulking. She was a widow with two sons, one of whom she told mother on a washing morning had been smitten with 'flu the night before and he was shivering so much she laid on top of him to try and get him warm.

Rather more by luck than athletic prowess she won a rabbit during the peace celebrations held on the village green in 1920. Four wild rabbits had been caught alive that morning by bolting them into nets put over their burrows, and kept in pairs in two ventilated boxes. They were released one box at a time into the middle of a large ring some thirty yards in diameter composed of anyone, man, woman or child who wished to take part, and if you caught one of the rabbits, not only was it yours but there was a money prize as well. The outcome of the first release was that one of the rabbits disappeared under Granny Iles full length voluminous skirt, whereupon she promptly sank to the ground clutching her skirt around her as much as she could to prevent it escaping. All this eventually led to considerable discussion as to whether Granny was the outright winner or half of the prize should be given to the young lad who pulled the rabbit out from under her skirt. It was all good fun anyhow and the other three rabbits got away. Another event I recall was the milk churn race over 50 yards, in which empty 17 gallon churns were propelled by gripping the knob in the centre of the lid in one hand leaning the churn over about 30 degrees and pushing it round on its bottom rim with the other. On a grass field surface this was quite exhausting.

Up to about 1930 the postal service to and from Baulking was carried out by postmen attached to the main Post Office at Faringdon, riding pillar box red bicycles equipped with a large metal parcel carrier over the front wheel and auxiliary canvass carrying bags attached to either side of the rear wheel. The outfit was illuminated in the front by a small paraffin oil lamp attached to a bracket on the hub of the front wheel; rear lights were not compulsory. There were two daytime deliveries and a separate evening collection from the post box which was by the side of the road, embedded in our garden wall at Church Farm. The postman I remember best was Jackie Lander, a kind, reliable and cheerful man, who never seemed to complain, come rain or shine he was always the same. Sometimes around Christmas time there could be half a dozen people standing in the dark near the post box waiting for him to weigh their parcels, which he did by torch light with a small vertical spring balance carried for the purpose. Having been weighed, the necessary stamps were affixed, the parcels secured to the front carrier by a cord in a pile sometimes almost too high for Jackie to see where he was going, and off he went on the five mile journey to Faringdon about two miles of which is uphill and he would certainly have to walk and push his load up Ringdale Hill. This particular run was known as the Uffington run and covered collection and delivery at Uffington, Baulking & Woolstone. The last named included a cottage at Woolstone Barn which could only be reached by a stiff climb up Woolstone Hill, followed by about a mile of cart track after crossing to the south side of the Ridgeway. An occasional early morning delivery to this address always brought a knock on our door later to say "Sorry I'm late Mam, but had to go to Woolstone

Barn this morning". I suppose the round trip for the three villages would be in the region of eighteen miles and sometimes the roster meant his journey had to be repeated in the evening for collection purposes especially during the periods of staff sickness. The advent of the Morris Minor motor car manufactured by Sir William Morris at Cowley in the early 1920's and its adaptation to small 5 cwt vans revolutionised not only rural postal services but many other businesses requiring economical methods of transporting light goods such as groceries and meat products to rural and urban customers alike.

What a famous man William Morris became and I recall the impact he made upon life in and around Oxford during the 1920's and 30's. His meteoric rise from a modest bicycle shop on The Plain in Oxford to setting up the mass production lines at Cowley producing first of all the bull nosed "Morris Cowley" and "Oxford" models to be followed by the "Minor" and "Isis" and many more.

In 1924 my father bought a "Cowley" from Ann's Garage, Faringdon whose showroom is now Dillons the Newsagents and Stationers shop. In those days it had two hand operated petrol pumps standing outside with swivel overhead booms allowing pedestrians to walk on the pavement underneath them while petrol was being pumped into the car standing close to the curb. Father bought the car for £199.10s. to replace a large Austin 20 which was too expensive to run and was not fitted with side-screens so on wet and windy weather could be a bit uncomfortable to ride in. The Austin 20 was the outcome of what was more of an arrangement than a deal with a well known character named Mrs Silver who lived at Warborough Farm, Letcombe Regis for whom father rode one or two point-to-point winners one of which being a horse named Plaudit on whom he won for her at Baulking on April 1st 1920. For this she presented him with an engraved silver cigarette case, now in my possession, which I treasure very much. To say that she was a character is a gross understatement. She hunted with the Old Berks, kept lion cubs as pets, had business interests in South Africa and was one of the first women to fly to The Cape, being the passenger in a two-seater bi-plane. She visited Church Farm one Sunday morning in her Austin 20 to say to father, "Cyril I'm going to Africa next week and I want your car to take with me. Austins are useless out there — the only makes one can get spares for there are Hupmobiles and Dodges, so I'll take your old Hup. "What about me?" said father "Oh you can have my Austin, its better than yours and I'll give you fifty quid for convenience sake if I can have it now — you can't refuse that and it will save me a lot of bother," was Mrs Silvers reply. So that more or less was that, the arrangement was confirmed over a "gin and ginger" and off she went with father's Hup which as a motor show model he had bought for £695 from an agent named Lieutenant Swain some two years earlier.

As I mentioned earlier, the Old Berks Hunt was then under the mastership of Dr Paget Stevenson. At that time he was anxious to bring a bit of uniformity to the headgear worn by the large number of farmers following hounds, so he made it known that he would appreciate it if they would all make an effort to turn out in velvet hunting caps instead of the variety of bowlers and cloth caps already being worn. It so happened one morning out hunting that father and Mrs Silver were sitting on their horses close to one another while hounds were drawing Rosey Covert. Mrs Silver was chiding father about his bowler hat, saying it was about time he conformed to the Master's wishes. Before father was able to reply, she knocked it off his head with her hunting crop and turned her horse round on it a few times trampling it almost out of sight into the waterlogged ground, laughingly adding "Now you will have to do something about it — won't you". Two days later her chauffeur delivered a hat box to Church Farm from Locks the well known hatters of St James Street, London, containing a velvet cap and a note saying "Hope it fits; if it does wear it — if it doesn't change it."

Quite recently Dr Paget Stevenson's Hunting Diaries came into the hands of Tim Weaving of Southmoor, the younger son of the late Guy Weaving, a great stalwart of the Old Berks and very close friend of Dr Paget Stevenson, a relative of whom thoughtfully decided to give them to Tim so that a record of that particular part of the Old Berks history can be kept in the area where it was made and to which it belongs. With the diaries are many interesting photographs, newspaper cuttings and letters, one of which was written to the Master by Mrs Silver after she had been reprimanded by him while out hunting the day before, for riding across a crop of winter beans. After apologising and promising never to do such a thing again she added a post script: "Why do farmers have to grow the ruddy things?"

*Mother's cousin, Cecil Selfe,
chosen to represent the
Southern Rhodesian Volunteers
at King George V Coronation*

*Granny Greenip
(née Mary Maria Selfe)*

Grampy (Ernest Liddiard) Competing in Faringdon Whit-Monday Sports
pre-First World War

Coxwell Villa, Faringdon. The house where the author was born.

'Sailor' boy, Ron, aged seven, with Cyril, his Father, and Kitty, his Mother

L.to R: *Kitty Liddiard (Ron's Mother), Ron Liddiard, Nellie Benson (Ron's Mother-in-Law)*

JUNIOR HOUSE circa 1923
Back Row; L. to R: Roberts I, Liddiard, Adams, Stafford, Wilkes
Middle Row; L. to R: Preston, Dickinson, Adie, Moore, Nunney, Hutchings I, Husband, Austin, Stokes, Brooke-Smith
Bottom Row; L. to R: Hutchings II, Forshawe, Cook, F. T. Brooke (Housemaster), K. A. R. Sugden (Headmaster),
Mrs. Parker (Matron), Wren, Tucker, Luker, Graham, Roberts II.

Ron Liddiard *Ann Liddiard*

WEDDING DAY
November 16th, 1939

RUBY WEDDING ANNIVERSARY
16th November, 1979

FORTY'S FARM: Flash Floods, May 26th, 1993
To the Rescue
Nicola (the author's granddaughter) and David Shirley (neighbour)

FORTY'S FARM: Flash Floods, May 26, 1993
Assessing the Situation
Pam (the author's daughter) and Jim Matthews (neighbouring farmer)
called to offer help.

STANFORD-IN-THE-VALE XI (Beaten 2-1 by Wroughton)
Final of Swindon 'Advertiser' Cup on the County Ground
Back Row; L. to R: *W. Smith (Trainer—wearing tie);*
Players: *Mattingley, Needham, Whittear, Collis, Godwin, Bowl, Liddiard*
Front Row: *Painter, Heyworth (?), Froud, Johnson*

Farming Club outing to London, Trafalgar Square, in the Mid 1950's
Back Row; L. to R: *John Florey, Derek Pike, Ivor Cooper,*
Jim Reade, Wilf Wyatt.
Kneeling: *Rob Pike, Ron Liddiard, Cyril Liddiard (Ron's father)*

ROYAL OBSERVER CORPS: Post X2, Uffington
Top Row; L. to R: *B. Packer, J. Bassett, J. Roberts, A. Richens*
Mid. Row; L. to R: *J. Willis, T. Weaver, R. Liddiard, E. Packer, R. Iles, B. Leahy, N. Grainger, W. Long.*
Seated; L. to R: *R. Williams, W. Packer, Officer in Charge of Oxford Group, W. Freeth, C. Field.*
Front Row; Cross Legged on Ground: *Jeanne Packer, D. Evans.*

The North Berks Herald & Didcot Advertiser 28 March 1941

The house next door to 180 Wootton Road, Abingdon, damaged by a bomb in an air raid in March, 1941.

155

Aerial View of Church Farm

Picking-up hay with a pitcher trailed behind waggon

Wheat Crop 1953 Yield 2 tons per acre

Author, driving the combine.

Photo: By F.A. Cox

Drawing tickets in Raffle at the Lurcher Show held at Peter Walwyn's Seven Barrows

Phil Drabble *(Wearing cap)*
Ron Liddiard *(with 'mike')*
"Bumble" Upton *(Hon. Sec.)*

Harry George and Cyril Liddiard (the author's father—on the grey)

North Berks Show, Faringdon House, circa 1924
Left: *Grampy—George Ernest Liddiard, Committee Chairman.*
Right: *Mr. W. N. Chambers, Auctioneer (Hobbs & Chambers), Hon.Sec.*

L. to R: *John Jeeves (Herdsman), Bill Liddiard (Son), Ron Liddiard (Father), CELIA (Star of Television Documentary: "To Celia and Son")*

Photo by:
Stan Hurwitz

*The author performing duties of Announcer at
Old Berks Point-to-Point at Lockinge on Easter Mondays*

*Cyril Liddiard, the author's father, being presented by Mr C.L. Loyd to
Queen Elizabeth, The Queen Mother, at Old Berks Point-to-Point, Lockinge,
Easter Monday, 1957.* **Centre Rear:** *H.R.H. Princess Margaret*
Others: *Bob Pike (Chairman—next to author's father); Mrs. A. T. Loyd
(far right); Inspector Coombs (Faringdon Police—holding horse rug).*

THE AUTHOR, RON LIDDIARD, CUTTING THE CAKE
at the 21st Birthday Party of the Faringdon Young Farmers Club

Golden night for YFC

Pictured at the Faringdon Young Farmers' Club golden jubilee dinner at Wantage Civic Hall on Thursday night are, from the left, chairman Rupert Burr, Tanner Shields, general secretary of the National Federation of YFCs, Peter Hobbs, president and Ron Liddiard who was the original founder secretary of the club 50 years ago.

Socialising at a Farming Club Ball at the Corn Exchange, Faringdon.
L.toR: *Ron Liddiard, Ernest Pearce, Jim Reade, Cyril Carter, Jim Kirby, Bernard Cook, Vic Tytherleigh, Rolly Maughan*

Final of the Berkshire Farming Club's Quiz at the Town Hall, Reading, Between Wantage (on the left) and Faringdon (on the right). Wantage Club were the winners.

Farewell Lunch, Caversham, on the retirement of J. W. Salter-Chalker, Rex Paterson & Ron Liddiard, from Southern Regional Committee of the Milk Marketing Board.
L. to R: *Ron Liddiard, Archie Lynham, John Farrant, (S.Reg.MMB Member), G. Peak, John Lunnon, John Salter-Chalker, Rex Paterson, Sir Richard Trehane (Chairman, MMB)*

Oxford Farming Conference, 1953
L. to R: *Charles Turner, Leslie Baker, Professor Harold Saunders (later knighted), Ron Liddiard, Norman Perkins.*

SIX
A Chink Of Light

It was early in 1933, at the age of 19, that I met someone who was to influence my future outlook on farming a great deal, enabling me to grasp the fundamentals of growing crops and feeding livestock, as well as impressing me with the importance of costing in relation to both. I think it is fair to say agricultural economics were not studied or applied by a very large proportion of farmers in those days. Farming was then accomplished rather more by "flying by the seat of your pants" and hoping everything would turn out alright in the end. There were of course 'born' farmers who were good at it but there were many others for whom it was a life-long struggle and who never seemed to be able to get out of the rut. Being well aware of this I felt any system by which at least some of the risk could be taken into account and budgeted for was bound to be an advantage and contribute a great deal towards future management and possible expansion. The man who opened this door for me was the late Freddie Cox who, with his wife Stella, became life long friends and I had the honour of proposing the main toast at their Golden Wedding celebration party.

It was in connection with the Faringdon Young Farmers Club that I first met him. The Club, incidentally, was the first Y.F.Club to be formed in Berkshire; had been inaugurated by a Captain O.W.Drew of the National Federation of Y.F.C.'s and as I had the honour of being its first Hon.Sec. he suggested I got in touch with the Berkshire County Councils' Agricultural Instruction Department at Reading for the purpose of obtaining speakers for our evening meetings. Grateful for the chance of a contact, the Chief Agricultural Organiser, Mr G.S.Bedford sent his Assistant Agric. Organiser F.A.Cox,BSc., to see me. After fixing up a course of six lectures on the Feeding & Management of Dairy Cows to be held at our headquarters, The Crown Hotel, Faringdon, and addressed by himself and various other well known experts from M.A.F.F. such as J.M.MacIntosh and Dr. Norman Barron from the Veterinary Department of Reading University, who told me some fifteen years later of a somewhat embarrassing situation in which he found himself while carrying out a similar task during the Second World War. It was on a foggy November evening that he left Reading to drive some thirty odd miles to deliver a lecture on the problems of fertility in the dairy herd to farmers and herdsmen assembled in the Village Hall at nearby Stanford in the Vale. Arriving in the blacked-out village and not knowing where the Village Hall was he knocked on a cottage door and was told it was only about a hundred yards further down the road: "You can't miss it, its almost in front of you." Having progressed the required distance by the dim light from his masked headlights he could see a large solid brick building with a heavy door framed by an arch. Being a bit late he gathered his brief case and an enamel tray containing part of the reproduction organs of a cow, pushed open the door and without looking at those assembled tip-toed quickly to the table at the far end of the room on which he placed his enamel dish and started to apologise to those present for being late, only to find all heads were bowed and hands clasped, whereupon to his horror he realised he was in the village Chapel and that the Hall must be a bit further up the road, so he crept out backwards closed the door quietly, sighed with relief and couldn't believe what had happened! I think he later left the academic sphere to take up a key post in commerce with Vitamealo, suppliers of high protein products for livestock feeding.

It so happened that my meeting with Freddie Cox coincided with an invitation to attend an Open Meeting of the Berkshire

County branch of the National Farmers Union to be held at Olympia, one of the largest meeting halls in Reading; for the purpose of listening to Professor A.W.Ashby extolling the virtues of the proposed formation of a Milk Marketing Scheme with statutory powers, enabling an elected Board of about twelve regional members and three special members representing consumers and government interests etc. to take control and market the whole of the milk produced in England and Wales.

My invitation came from Freddie Poole, a family friend, who farmed at Field Farm, Longcot. He was a staunch supporter and past County Chairman of the N.F.U. He had also shown great interest in the formation of the Faringdon Y.F.C. and was the Club's first President and Leader. Coming into farming from a business background and being involved at County level in farming affairs, he was well aware of the bankrupt state in which farming, as well as almost every other industry, was in.

So while at supper one evening at Church Farm, he told father it was essential that I should hear what Prof. Ashby had to say, as he thought it was about the only way in which stability could be put into dairy farming and that once that had been achieved the rest of farming would benefit a great deal as well. It was the first really important meeting I attended and I recall being transported to Reading in Freddie Poole's cosy little French Dion Bouton two seater coupe. There were over 700 farmers present, A.H.Cornish of Abingdon took the Chair supported by his Vice-Chairman "Jock" Simmons of Hampstead Marshall, both very good fairly large-scale farmers. Professor Ashby gave a long, excellent address and I had the impression that every word was listened to with intense interest. Many questions were asked and answered and a show of hands at the conclusion of the meeting, indicated that almost to a man, everyone was in favour of adopting the scheme when it was finalised and presented for acceptance. I think it is fair to say history records that Freddie Poole was right and by and large the Milk Marketing Board has served its purpose admirably since it was formed sixty years ago. It is apparently now looked upon as a monopoly and has had its statutory powers withdrawn and converted into a voluntary co-operative organisation known as Milk Marque. While accepting that conditions have changed since the introduction of milk quotas, I find it difficult to accept the possible fragmentation that may occur within the industry; I fear that interfering with an organisation that has been such an outstanding benefit to producers and consumers alike for so long, may result in a return to the difficulties which previously existed, for which the Board was specifically set up to deal with in the first place.

My involvement with "the two Freddies" at more or less the same time revealed a fresh line of approach and purpose and I began to realise there was a great deal of help available in the form of agricultural research and that it could be of great benefit to get to know and believe in what had already been discovered and endeavour to put some of the ideas into practice. There is no doubt it helped to fill a gap in my education which might have been provided by attending an Agricultural College, but they were few and far between, quite expensive and never even considered. Perhaps a little unknowingly "the Freddies" inspired me and kindled a desire to seek advice, whether by listening, visiting other peoples farms, attending or taking part in demonstrations on changing techniques etc, all of which lead eventually to co-operating voluntarily in cost recording schemes relating to milk production and general farm accounting, operated by the Agricultural Economics Department of Reading University as well as by I.C.I. (Imperial Chemical Industries) the latter being more concerned with grassland management and its financial returns relevant to the use of fertilisers. This has always been a road down which I have been happy to go for the last sixty years. The outcome being that I have met scores of wonderful people prepared and willing to give me of their time. I have not made a fortune but with their help and God's blessing the family have all slept in the dry and had something for breakfast each morning.

Returning to 1932, following a suggestion from father I decided to expand my small pig enterprise and dip my toe into the pedigree pool by purchasing a registered Tamworth sow, at a draft sale from the Burnham Herd belonging to E.Clifton-Brown Esq., of Burnham Beeches who was then Speaker of the House of Commons and a brother of Vice-Admiral Clifton-Brown, who lived at Stanford Place, Faringdon and was a keen advisory member of our newly formed Y.F.Club. The sow rejoiced in the name of Burnham Mercy and was due to farrow in 3 weeks to a mating with a boar that had been twice breed champion at The Royal Show. The Tamworth breed are sandy in colour and primarily a bacon type pig but mated with a Berkshire produced a very attractive looking sandy and black spotted offspring, very suitable for the pork market and sold well as weaners, which was my main objective. Disappointingly, Burnham Mercy only produced a litter of three, made up of two boar pigs and a gilt. The latter I mated at six months old with a further purchase of a boar from the Burnham herd named Burnham Brutus who also had some of the same championship blood three generations back in his pedigree. The result of all this was a litter of five boar pigs and four gilts. I decided to breed from the gilts and rear the boars for the purpose of selling them at six months old in one of John Thornton & Co.'s collective pig sales at Reading Cattle Market. It was a condition of sale entry that each boar should arrive in a two wheeled wooden crate to facilitate handling. W.Packer & Sons of Uffington made these for me and painted The Baulking Herd of Tamworths with my name underneath on them, which made me feel I was really in business! They cost £2 ten shillings each.

The catalogue contained close on a thousand pigs of all breeds, genders and ages and the auction was shared by the well known brothers Neville and Frank Matthews. In order to give the boars time to settle down I took them to Reading the day before, kipping down on some straw near to them during the night. Due to the large number of pigs forward it was an early start to selling the following morning. Trade was very slow, with the auctioneers having to work hard. Despite my entries being the only ones of the Tamworth breed on offer, I was feeling very despondent when Frank Matthews came to the pens and told me that Mr C.J.Twist the agent for the Burnham Estate, from whom we had obtained our original breeding stock, was interested in buying the boar, registered as Baulking King for export to Australia. He also said that as he would not be coming under the hammer until late in the afternoon Mr Twist could not stay to see him sold but would go to eight guineas for him. This pleased me a great deal as most of the boars from other breeds were struggling to reach between four and five guineas. "Baulking King" was of course knocked down to Burnham Estate but despite putting him through the ring as the first of my consignment, the price of 8 guineas paid for him failed to liven up the bidding for the other four which averaged a miserable three guineas a piece.

The reason for mentioning this story is that about five years later George Heavens, the Faringdon butcher showed me a report of the premier Australian Show at Brisbane which someone had sent him and which recorded that a Mr Somebody from the Adelaide area had taken the supreme championship with a Tamworth boar which was sired by that famous boar "Baulking King" with whom he had won many prizes, since importing him from England. It was nice to know the name of our tiny village had found a mention on the other side of the world, but long before I had heard about it, the bottom had dropped out of the pig trade and what had started as a small sideline had begun to take up most of my time and didn't mix with plans that were beginning to take shape in my mind concerning possible expansion of the dairy herd. It was an old saying that pigs are either copper or gold so having had a tiny share of the gold I thought it best not to have too much of the copper and sold out completely in a single deal to the late Bert Haynes who then farmed at West Mill, Watchfield and had started doing a bit of dealing as well. He later moved to East Challow where he started a scrap yard which developed into the sizeable business it is today, buying and selling all manner of goods and equipment. I can well remember feeling very sad helping to load the whole lot, big pigs, little pigs, pigs in between, pedigree and cross breeds alike into his lorry which had to make two journeys and not being able to find the courage to watch the last one go out of the farm gate. Something I have learned in my farming life is that animals are inevitably a bit like human beings, "you love 'em all but some very much more than others."

I had enjoyed developing my small pig enterprise and learned a little from the various characters, loosely know as "pig-pokers" when trying to get the best possible deal. I also enjoyed taking pigs to market at Faringdon, by horse and milk float. The latter being a converted horse drawn, two wheeled ambulance cart used in the 1914/18 war. It didn't need much converting other than removing the white canvas hood with the large Red Cross on each side. Father bought it at a Government Surplus sale. I often wondered what poignant scenes it may have witnessed. Sadly very different I fear to when I was transporting nine porkers in it to Faringdon Market. Half way up Ringdale Hill, the gradient was sufficient for all the pigs to slide backwards in a heap against the tail board and break the cord securing the pig net with which they were covered to prevent them jumping out. They didn't all fall out ,but the four that did scampered off back towards Fernham from where we had come. I managed to secure the rest and trotted back down the road after them, overtaking them to turn them into the yard belonging to "The Woodman" public house, from which emerged a very kind lady named Mrs. Warner, the licensee, who with a broom in her hand and a lot of chasing about helped me to usher them into an adjoining stable, from where I was able to catch them one at a time and lift them back into the cart. As they weighed around 80lbs apiece, it wasn't easy — even worse in fear of missing the market I hadn't the time to stay and quench my thirst. Despite everything, a few sympathetic words to likely buyers by Mr. Chambers the auctioneer helped towards a satisfactory result. So with a song in my heart, Chang (the name of the chestnut mare with a big white blaze) and I returned to Baulking having broken our journey to say thank you again to Mrs. Warner for her kindness and offer her some remuneration which she stoutly refused to accept even though the spotlessly white apron she was wearing when helping me earlier was far from the same colour when the pigs were finally loaded.

Able now to concentrate more on the cows, I was encouraged a great deal by the outcome of Freddie Cox's suggestion that changed our feeding methods by weighing a few random trusses of hay cut from a rick so that each cow could be given a daily allowance of around 20lbs as a maintenance ration on top of which she would receive a production ration of 4lbs per gallon of a mixture of straight feeds consisting of crushed oats, flaked maize, linseed cake, decorticated ground-nut cake and a very small amount of fish meal usually never more than 5%. To this was added a very cheap mineral supplement consisting of ground chalk, steamed bone meal and common salt, in equal parts. This was all mixed on the barn floor by turning the initial heap three times with barn shovels which were lighter and larger than ordinary shovels.

Half of each cow's daily production ration was weighed and put into a bucket with her number on it and was put into her individual manger immediately prior to her being milked. Cows are more relaxed when feeding and this contributes towards what is known as milk 'let-down'. This procedure was repeated after morning and evening milkings and meant that a cow producing 5 gallons per day, the latter amount rising or falling in relation to the amount she produced on the weekly day on which each cows milk was weighed and recorded both for feeding and longer term breeding purposes.

This controlled method was based on the amount of starch and Protein Equivalent necessary to keep a cow alive and fit, plus the amount necessary to produce the milk she is yielding. As a cow can only consume around 33lbs of dry matter per day it is necessary to provide part of her ration in a concentrated form, using the nutritive digestible values of feeding stuffs available to produce a palatable and economical balanced ration. This replaced a much more laborious method of indoor feeding based on growing a root crop, usually mangels which were harvested in the autumn and stored in a long clamp either in the fields in which they were grown or close to home. To keep the frost out, the clamp was covered with what was called rowan. It was a collection of rough grass and weed trimmings etc. cut by hand from the banks and hedgerows surrounding the arable areas before they were ploughed for the following crops. Harvesting the mangels or 'mangling' as it was often named was quite a performance. It was usually done in early October but could be a bit of a long drawn-out affair if the ground became rain sodden. The mangels, weighing somewhere around 4lbs apiece, were pulled out of the soil by gathering up the leaves in your left hand, lifting it while putting an old carving knife blade behind the bunched up leaves, cutting them off at the same time as throwing the mangel into a small heap; leaving yourself with a handful of leaves. The leaves were then thrown in a rough circle around the heap and used to cover the mangels up in case of frost before they were picked up by horse and cart to be clamped. A heap in the field consisted of the crop from about 80 square yards, there was no point in trying to throw the mangels too far and it was a back-aching job anyway.

It was a common sight during the autumn to see many small acreages around the countryside dotted with small heaps of roots about 16 yards apart, waiting to be carted in order to clear the ground in preparation for the following crop usually winter wheat, which was a good bet if you could plant it in favourable conditions before the middle of November. I remember we were all very chuffed to grow our first two ton crop of wheat in this way, sowing it on Guy Fawkes Day in a five acre field named Little Maids Mead from which we harvested 98 four bushel sacks of wheat, equivalent to 44 cwts per acres. It was cut with a binder, stoked, carted, ricked, thatched and eventually threshed by "tackle" belonging to larger farmers such as Harry Baylis from Hatford, or Raynard Adams of Fernham, who visited farms in the neighbourhood as threshing contractors supplying a team of three operators, consisting of a foreman who set the tackle up and looked after "sacking-off" of corn, a feeder who cut the string bonds on the sheaves as they were thrown to him by prong from the rick, and woe betide anyone who threw them to him other than the way and pace in which he wanted 'em!. The third man dealt with the discharge of the threshed straw either into a machine which tied it into bundles or a wire tying high density baler. Sometimes it was just discharged into an elevator for the purpose of storing it in a rick for thatching or litter usage.

A good crop of wheat cut with the binder usually provided a quality of straw suitable for thatching houses and cottages and we used that from our two ton crop to re-thatch our cottage by the Church, which had been converted under the Housing of Rural Workers Act from a three family dwelling into two. A grant available to assist in the development, also imposed conditions that insisted on the accommodation being occupied solely by rural workers and that the rent charged should be no more than three shillings and five pence per week, and that such restrictions should elapse after twenty years. The cottage was thatched entirely and beautifully by Harry Whitfield, a tall, upright man from Stanford in the Vale. He did all his own "yelming" which consisted of drawing the straw from a lightly shaken up pile of straw which had been previously sprinkled with water. With a kind of over-arm action the original two handed pull of straw from the pile was added to by gathering to it, grabbing a bit more on each over-arm action, then it being already reasonably straight, throw it on the ground, tidy it up by raking it through with his fingers, taking any rubbish out before placing the amount which was about a medium sized armful into an ash fork about four feet long cut from coppice, known as a "yelm" carrier. When this was full, it was secured across the top with a rope loop and weighing about cwt was then ready to be carried up the ladder on to the roof for laying.

It was fascinating watching him make the laths and spelks out of lengths of withy wood about as thick as one's wrist, obtained for thatching by pollarding withy trees after about three years of growth. Harry would sit on a low stool and having split the spelks to the required size he would place them across hard leather knee protectors worn primarily to save his knees from getting sore on the ladder rungs; he would sharpen them with precisely three cuts with his bill-hook, "two long and one short" he would say, and "the bill-hook needs to be sharp enough to shave with". He would then bend and twist them in the middle making them into a large wooden staple strong enough to push into the thatch and hold the straw in position in the strongest of gales and secure the laths that run along the eaves. I remember him saying after he had finished covering his fascinating handicraft with mesh wire netting that it would "last another thirty years but you might have to patch up the ridge after twenty" — and he was right.

From mangels, wheat and thatching I return to describe the method of winter milk production being practised by the majority of farmers at that time. It was based mainly on the feeding of bulk foods such as hay and roots, which were usually mangels. These were pulped and mixed with wheat or oat chaff, helped down at times by the addition of some molassine meal, a by-product of the sugar industry. The hay was fed at the rate of 'a good armful every night and morning'. A bowl or two of a brand of compound cake at each milking time, depending on what it was 'thought' the cow was yielding.

Daily or weekly recording of the yield of individual cows was in its infancy and not being used significantly by the majority of farmers, so output was measured by taking note of the total daily gallonage put on rail or delivered to the dairy depot. In this way, to all intents and purposes, the cows were fed on a herd basis and there was no doubt that milk output relied to a large extent on the quality of the hay made during the preceding summer. I have heard it said more than once, that the only supplement needed by good hay was a few cigarette cards (which were free) but poor hay needed a cheque book.

There were those too, mainly of the older generation, who looked upon mangels in the same way as others that believe 'an apple a day keeps the doctor away'. One such being asked while he was carting mangels home to store them for winter feed, "Why he was carrying so much water about?" (mangels were about 80% moisture) to which he replied "Ah, but this is GOOD water".

Be that as it may, with father's blessing we embarked on individual feeding with balanced rations for the winter of 1934/5 with the result that our milk sales for that year from the same number of cows rose by 36% over the previous year. While that might sound a bit like trumpet blowing it's nothing of the kind, it just proved how badly we had been doing it and that there was help available to bring about improvement.

Fred Ridley, a character with a great sense of humour who lived at Uffington, reared a calf of two, kept goats and hens etc. and as previously mentioned was involved in ejecting gypsies from Wentworth's field in Uffington, happened to be working for us at the time of our change in feeding policy. I remember him saying to father while getting up from a cow to empty his partly filled bucket and return to his milking stool to finish her off.... "I reckon you'll have to give us another shilling or two Guvner for getting all this extra milk out of 'em" which father countered by saying, "Ah, but don't forget the times when you sat down there Fred and didn't get half as much".

Which reminds me of the old smallholder George, who returned from market having purchased a replacement for one of his herd of three who had become sterile through old age. Having unloaded and chained his new purchase up safely by the neck and had a cup of tea, his wife, Liza, said you better go and ease 'er a bit, as no doubt she'd been stocked up for sale and hadn't been milked out properly for 24 hours. So off he went across the yard to his small cowshed, hung the lantern on a nail in the old beam and got on with the job. About three hours later Liza was worried as the old man hadn't come in for his supper so she went across the yard to the cowhouse to see if he was alright. To her amazement there was a line of buckets full up with milk standing along the wall at the back of the shed and old George sitting on his three legged stool with his head tucked hard into the new cow's flank, still milking away like mad, prompting Liza to say that "it looks like we had got a bargain". "Arr", said George without a turn of his head, "and 'ers still gaining on I".

While on the subject of smallholders and their cows. Grampy Ernest loved telling the story about another Ol' George — a smallholder who had somewhat mysteriously lost a cow and despite reporting the matter to the police, not a vestige of information concerning her whereabouts had come to light. About a week after her disappearance the vicar met George in the village and inquired kindly if he had news of his lost cow "No Sir, no tidings at all; can't understand where her can be," and followed up by saying "S'pose you couldn't give it out in Church on Sunday — might prick somebody's conscience, if you know what I mean". "Well George, I'm not sure if I can do that, we only see you at Harvest Thanksgiving and Christmas". "Right enough" replied George "but I'll be there and promise to come a bit more reglar if you could help I". "Oh alright just this once then" says the vicar. "Thank you Sir, I'll be there and as you knows I be a bit deaf so shall be in the front pew" which clinched the deal. Sunday morning arrived and George cupped his hand to his ear as the Vicar climbed into the pulpit and proceeded to publish the bans of marriage between two of his young parishioners whereupon George got the impression he was giving out the notice about his ol' cow, and when the vicar asked if anyone knew any just reason or cause why they should not be joined together in matrimony Ol' George raised himself to his feet and interjected, "I forgot to tell ee Vicar 'er only got one teat".

It wasn't long before our own increase in production from existing resources led to thoughts of increasing herd size and improving winter housing conditions by closing the open fronted cowsheds, doing away with the old wooden yokes, installing metal divisions and neck chains, for the purpose of individual tie-ups, each with their own drinking bowl, which would replace drinking from the pond in the corner of the cowyard during the winter, rich no doubt as it was in intrinsic values. It was obvious that such expansion would require a certain amount of extra labour so the installation of a milking machine became a distinct possibility. However, all these bright ideas were put into proper prospective when it was realised that the first priority was an ample supply of good clean water, when at that time all we had was a well in the dairy from which water was pumped by hand into a 400 gallon tank situated in roof space above what was once the old cheese room, having been built on to the farmhouse in 1822 as recorded to this day on a chalkstone built into the red brick wall at the eastern end of the addition. The water was unfit for human consumption containing an unacceptable amount of ammonia, probably due to its close proximity to the cow-yard over so many years.

Pleading with the Faringdon R.D.C. to allow us to use the recent piped supply to the village which passed the farm gate, they said conditions attached to the scheme that water was available from standpipes and for domestic use only must be strictly adhered to, despite our obvious offer to purchase a metered supply. I'm sure too, others in the village would have welcomed the chance to do the same. Alas it was not to be, so it seemed everything was lost until Freddie Cox came up with the only other possible solution; that of sinking a borehole which he warned could be expensive, but if successful would not only allow the desired expansion to proceed but would increase the value of the property. Having fairly recently obtained his B.Sc.Agric. at Reading University, he was still in close touch with many of the departments including that of geology, headed by Prof. Hawkins, who very kindly researched the possibility of obtaining a supply of water sufficient for our needs. Coming to the conclusion that it would be advisable to bore to a maximum depth of 80 feet, he anticipated that with all the local information we had been able to give him concerning the depth and water levels of existing dug wells in the parish, some of which we descended by tying ladders together; that we could possibly strike a supply somewhere between 50 and 60 feet.

Armed with this information we contacted Guthrie Allsebrook & Co., Water Engineers of Reading (also well known to Freddie) and they gave us a quote to put down a 4½ inch bore at £1 per foot and should water be found they would provide a further quote for lining the bore with steel tubing inserting a pump at a suitable level with a rising main and the necessary head gearing to operate in conjunction with a 3 h.p. petrol engine.

Without the engine we reckoned this could increase the overall cost to a little over £200, which caused father to remark that we hadn't got the money and with all the other items, milking machine and the like he couldn't see where it was coming from. I think it worried mother a bit too, and they were both unaccustomed to being offered advice, by someone they had not known for very long, at the same time, thinking perhaps that I was a bit too gullible.

As far as I was concerned I trusted Freddie implicitly, looked upon him as a real thinker and was full of enthusiasm for what he had to say. Although I think he knew that he was not the best of disseminators, he more than made up for any shortfall in that respect by his absolute sincerity and genuine desire to help.

Aquainting him with the position he had a word with one of the Allsebrook brothers who said they would be prepared to help us by hiring out the boring tackle consisting of about fourteen 6 ft steel rods a couple of T chisels, a debris recovery shell etc. for 2 months at a cost of £18.

Although this was a great financial saving I was more than apprehensive about the practicality of actually doing the job. However Freddie who had married Stella Killick, daughter of a Kintbury farmer, and a former fellow University student, said he was prepared to oversee the operation by showing us what to do, start us off and come from Tilehurst to stay with us for a few weekends with Stella, to assist and monitor progress. This offer was gratefully accepted on the understanding that whatever the outcome they would allow us to recompense them in some way or other.

So it was in 1934 that, having decided it would save time if we commenced boring in the bottom of the existing 30 ft well in the dairy adjoining Church Farmhouse, part of the tiled roof was removed to allow a tripod rig of three scaffold poles to be erected over it to carry a pulley, over which a rope would travel when lifting the boring rods up and down, having first been used for lowering 30 feet of 4½" steel lining into place in the middle of the well. This was pile driven as far into the bottom of the well as possible to seal the borehole from the water already in the well and secured at the top of the well with a wooden clamp embedded in the well wall, for the initial purpose of maintaining a rigid guide for the rods to travel up and down in and later to hold the rising water into which the pump barrel and rising main would be suspended.

The lay-out for the boring operation consisted of our portable 3 H.P. Lister engine driving an old fashioned low geared winch which Freddie had spotted in a bunch of stinging nettles while visiting a farm on his rounds near Reading. The surface of the winch drum was well greased and a strong rope was wrapped round it about four times, with one end running over the pulley hoisted over the top of the well, down to the boring platform at ground level; while the other end was held firmly in one's hands, pulling it hard to make it grip round the drum, and so lifting the boring rods and chisel about 2 feet then suddenly letting the rope go, so that the full weight of the rods pounded the chisel into whatever type of soil was being encountered. This pulling and letting go procedure was accomplished from the seat of an old kitchen chair, rather like rowing a boat. After a while it was quite easy to settle into a rhythm and it was satisfying to see the chalk marks disappearing. These were made on the rods before each boring session, which aimed at a depth of around 2ft 6ins. When this had been reached the rods were withdrawn by a longer hand overhand pull on the rope so that they could be unscrewed in six foot lengths, once a steel shoe had been placed under the collar on each rod to prevent those below ground level slipping back down the bore hole. Which of course would have been disastrous.

As soon as the borehole was free of rods, a debris shell consisting basically of a 4 inch steel tube with a metal hinged valve on the bottom was lowered by screwing the rods up once more dropping it on to bottom a few times, forcing what had been bored into it which was held there by the foot valve closing immediately the shell was lifted and taken to the surface by what seemed at the time a never ending procedure of screwing and unscrewing steel rods, but again it was interesting seeing what we had removed from the bowels of the earth. We did in fact put a sample from each shell on the top of a nearby wall. In the knowledge of what was discovered below ground in the immediate area some 50 years later, it is possible had we thought of getting some analysis done, the presence of calcium montmorillonite (Fuller's Earth) might have been discovered earlier. Personally, I'm glad it wasn't.

At one stage in the operation having reached an overall depth of about 53 feet, quite suddenly after passing through one or two thin ironstone bands and a great deal of dark grey mudstone we hit what felt like solid rock which caused the steel rods to bounce and shudder and made the handles clamped to them for the purpose of rotating the cutting edge of the chisel in a clockwise direction between every strike, very difficult to hold.

A phone call to Freddie reporting that we had been bashing away for three days and had only managed to get a couple of inches progress brought little cheer. Apart from saying he would have a chat to Allsebrook concerning the possible use of another type of chisel head he said "keep going, Stella and I will be down on Saturday afternoon for the weekend". When they arrived we had still only managed another couple of inches. While I was looking in vain for any further piece of helpful equipment in the boot of his car, he said "All you will find in there is a large helping of patience which is necessary when doing a job like this"!

As if to prove his point, a further half an hours pounding and we broke through the 6 inch seam of ironstone as suddenly as we had struck it almost a week earlier. It was after this that the chisel and rods came up much cleaner and there was much more sand in the debris, which we decided might indicate that water was coming from somewhere around this depth. However, we drilled on to our intended target of 80 ft, rigged up a temporary pump capable of throwing out 350 gals per hour to see what happened. The result being that the water level in the borehole dropped 7 ft quite rapidly to a point where it settled and remained for 72 hours of continuous pumping.

This gave us enough confidence to set the whole thing up permanently, so we withdrew the pump and rising main and proceeded to line the whole borehole by pile driving the steel lining tubes, some of which were perforated to allow water access. All went well until the bottom liner reached the seam of ironstone that had caused us so much trouble earlier at around 53 ft and we realised we had made the mistake of drilling the borehole with a 4 inch chisel and although the inside diameter of the lining tubes was only 4½ inches the outside diameter was 5 inches and while we had been able to punch the liners through clay and mudstone, trying to get them through a 6 inch thick seam of ironstone was like trying to get a pint into a half-pint mug. With all our effort being of no avail, we took the chance of leaving things as they were and hoping we would be able to get the lifting main attached to the bottom of the pump barrel far enough down the unlined part of the borehole to pick up the water as we were still unsure as to the depth from which it was coming. Assembling the pump line to reach down to 70 ft failed dismally when it would not drop down any further than 60 ft due to the unlined borehole filling with sand to this level, which was evident from the contents of a debris shell, dropped down to see if it would provide any useful evidence. In consultation with Prof. Hawkins and Messrs Allsebrook it was assumed that an ample supply had been found in the region of the ironstone seam. Our 3 day test pumping had also removed an appreciable amount of sand with the water which had formed a cavity probably just below the ironstone seam and, if we aimed to pump from this area, they had reasonable hopes that we might be lucky, although a small amount of sand in the water might be a problem until the operation settled down. How right they proved to be; we set the outfit up by bolting the pump head gear unit and the Lister petrol engine to two halves of a sixteen foot steel barn stanchion, sawn through with a hack-saw. We connected the 1½ inch rising main to the old hand pump in the corner of the dairy by taking out the tap. We increased the size of the overflow pipe from the header tank in the roof and Freddie finished the job off by graduating a white painted gauge board, fixing it on the apex of the end wall on a level with the tank, making an arrow shaped marker attached to the weight from an old set of grass harrows which attached to a light chain moved up and down the gauge indicating the amount of water in the tank. It was controlled by the use of a weighted float made by sawing the handles off two old fashioned galvanised hand basins, bolting them together and soldering the rims. By floating this on the water with the other end of the chain attached to it and running over a small pulley wheel above the tank, it worked like a dream. The total cost turned out to be £78 plus a weeks holiday for Freddie and Stella at Bournemouth. I remember feeling very chuffed that Grampy Liddiard, who was suffering a great deal from arthritis, paid us a visit to see the boring operation and again when we had completed the job. His "Well done" meant a lot, as mother reminded me he was not particularly generous in bestowing such expressions of admiration. I was grateful to Harry Eltham who was my workmate during week days who said he was only pleased to help as it would relieve him of that "ruddy job of hand pumping seven days a week".

We were now able to proceed with improving winter housing conditions for the cows which had the desired effect on their yield and well being. As the water supply was perfectly adequate during the winter not only for the cows but most of the young stock as well, it was decided to rear a few more heifer calves to help increase the herd size later and as a more immediate step to buy half a dozen freshly calved heifers during the coming autumn, by which time I was hoping to have persuaded father into buying a milking machine and reduce the number of men we employed from four to three.

Machine milking in those days was looked upon with a certain amount of fear and apprehension. There was a lot of talk of ruining the herd with mastitis, cows not letting their milk down and having to strip them out by hand following the machine — so what was the saving — etc. etc. It so happened I had been playing rugger with a chap at Faringdon named "Curly" Dryden who had just come to live at Buckland and was a sales representative for the milking machine makers Alfa-Laval of Brentford, Middsx. Being impressed with the literature he gave me, "Curly" suggested I get father to go with me to see a machine he had just sold to Jack Barrett at Manor Farm on the Lechlade road just outside Faringdon. Jack had only been using the machine a few months and his opening remark to us was "I'm no mechanic so when I saw all the bits and pieces unpacked and laying all over the dairy floor I told 'em I should never know what to do with it all and they had better take it away. However," he went on, "I'm glad they didn't. After a very few milkings the cows took to it and so did we — wouldn't be without it. Mind you I shouldn't want any fool to use it but I'm lucky I've got a very good chap".

I think father was quite impressed by what he saw and heard

but was worried as to where the money was coming from. Curly had quoted us £134 to install a 3 bucket unit with a vacuum pump, and overhead pipe line sufficient to accommodate 38 cows.

Anxious for a deal, he asked me what I thought about mentioning hire-purchase to father as a possible solution. I said, "try if you like but he'll probably 'blow his top'." However Curly took a chance, saying that, "all it will cost you is £6 three shillings and four pence a month for two years," sugaring the pill by adding, "we will put it in for you to try for a month and if you don't like it we will take it out and it won't cost you a penny". "Come on Dad," said I quickly. "That's got to be a fair offer". After a lot of talk about never having had anything on the 'glad & sorry' before etc., he agreed and in two years time we gave Alfa-Laval a shilling and they took the small brass plate off the vacuum pump stipulating that they had been the owners of the equipment up to then and it was now ours.

Curly Dryden, named so as he had unfortunately lost his thatch as a results of alopecia became quite a character later on. When war broke out in 1939, he took advantage of the R.A.F.'s offer of a Pilot-Officers' rank to those who enlisted immediately as air-gunners. He rose, I think, to the rank of Squadron Leader following a long stint of missions, one of which left him drifting in an inflatable raft in the Mediterranean for two or three days before he was rescued. Being subsequently 'grounded' for a while he spent a period as O/C Shellingford aerodrome which was an initial pilot training unit using Tiger Moths. Story has it that following an emergency landing by a trainee into a very small field close to the drome, it was thought the only way to get the plane which was virtually undamaged, out of the field was to dismantle it and take it a few hundred yards down the road to base by lorry. Upon learning of this suggestion, Curly took it upon himself to visit the scene to see if there was any other option. It was said that after sucking his finger and sticking it up in the air to confirm wind direction he taxied the biplane to the furthest point in the field down wind, got his lads to turn it round and pull it backwards as close to the hedge as possible and hold on to the tail plane while he revved the engine and waved his hand from the cock-pit to let go. The result of this being that everyone present held their breath as he managed to clear the hedge at the opposite end of the field by inches, do a couple of circuits and put the Moth back down on the drome. Shortly after the War Curly made quite a mark in the world of motor racing, while at the same time fulfilling the role of mine host of the 'The George' at Dorchester-on-Thames. Sadly, at too early an age, he lost his life while competing at Castle Combe.

I don't remember father ever having bought anything in that way before or afterwards. If money was needed to finance improvements, and it often was, then within reason the bank was always found to be sympathetic. I always felt basically it was better to have a single debtor than fragment one's borrowing by taking advantage from other sources such as agricultural merchants who were sometimes prepared to assist by extending credit facilities. I once heard a well known farmer say it was very much easier trying to keep one wolf from the door than a whole pack of 'em. He was a Scotsman named William Alexander who farmed at Ensford in Kent. While visiting his farming enterprise as a member of the Filkins & Faringdon Farming Club not long after I was married, I had the experience (and honour) of sitting next to him in an old Ford Prefect while he piloted us around. It was during May, the blossom was out and the Garden of England was looking at its best. As we travelled along one side of a valley I remarked what a beautiful sight a particular farm presented nestling at the bottom of the valley with its 3 or 4 Hop-oasts, appearing almost like a picture on a calendar. "Funny you should say that" he said with a cigarette dangling from his top lip, "I bought that holding last back-end; hadn't got any money; but I managed it somehow". "That's particularly interesting to me as a young man with a wife and small family," I said, "trying to squeeze a reasonable living from a few cows on 180 acres in Berkshire and I have always wondered how some people are able to do things without money?" "Well lad," he said slowing the car down a bit, "I'll give you a bit of advice. When you want to borrow from the Bank, it's no good going cap in hand, you've got to go in the 'sweat box', tell the bank manager exactly what you want to do and before he can even start to answer, you say 'and I want to give you the first chance to lend me the money to do it'." — I admit to reciting the story to one or two Bank Managers for fun but have never had the guts to put his advice into serious practice!

While visiting one of Alexander's dairy units during afternoon milking I was more than surprised that one of the two lads who were doing the milking turned out to be a cousin of mine, Derek Holtom, who lived in Canterbury and was doing a stint at Ensford as a farm pupil, following his marriage to Mollie Sale, the daughter of a well-known hunting character within the Puckeridge country. He went to farm at Clothall Bury near Baldock and has now retired to the quieter surroundings of Exmoor, where he can enjoy watching hounds work during the fox-hunting season.

Based at Canterbury the eight other farms we visited on this trip provided a variety of interest from the specific sheep grazing techniques practised on the Romney Marshes, employing the services of experienced "lookers" whose job it was to assess almost daily the desired rate of stocking in sheep per acre, necessary to preserve the quality and quantity of grass available to benefit both sward and sheep. In order to achieve this sheep were drafted in and out of these prime grazing areas from holding paddocks.

This contrasting greatly with corn growing on the Isle of Thanet and the fruit growing areas nearer Canterbury. Apart from Alexander the other farming family names I remember are Fyn-Kelsey, Mount, and Quested. A member of the last named I recall as not only being a farmer, he was a butcher and above all mine host of the Fleur-de-Lys at Sandwich where lunch had been arranged for our party of around a couple of dozen on the Sunday, the afternoon being free for us all to do our own thing. During the pre-lunch session in the bar Robert Henly who farmed at Harn Hill near Cirencester and married a distant cousin, Mary Chillingworth, came to me quietly and said he had discovered there was a fruit machine on the premises paying out in real money which was then illegal except under certain licence conditions. So we armed ourselves with a couple of pounds worth of sixpencees (40 each) and he led the way upstairs on to a landing where he opened a door with LADIES on it to reveal a very small room with the gambling machine attached to one of the walls. We turned the key on the inside of the door and began trying our luck. As usual we slowly began to lose more than we won and eventually lost all our original £2 stake money. As we opened the door to leave we were confronted by a stranger who looked at us sternly and said "Gentlemen, you have been gambling on the machine which you probably know is illegal and as it is Sunday as well you are contravening the laws concerning gambling on the Sabbath on licensed premises. I have been waiting sometime for this opportunity, so must ask you both for your names and addresses." Robert and I looked at one another in muted astonishment but before we could think of anything to say, the chap burst out laughing and said, "I expect you've lost what you brought up here, so come out of the way and let me have a go I should think the jackpot is about ripe"!! From what the landlord told us after lunch on this occasion it wasn't but this particular "regular" was a very shrewd observer of people who obtained change over the bar and disappeared upstairs.

Two years after the installation of the milking machine and the accompanying herd expansion it became necessary to produce more hay for winter maintenance. An immediate boost was gained by sowing down our small arable acreage to a grass and clover mixture recommended by Freddie Cox and known as the well proven Cockle Park mixture consisting of Ryegrass, Timothy and a small amount of Cocksfoot (not particularly suited to heavy soils but included to see how it performed) some late-flowering Montgomery red-clover and 1½ lbs of Kentish wild white clover per acre. This step proved to be quite revealing as it became established. Later on, employing a stationary baler we weighed the output of hay from this area which yielded 56 cwts per acre and did the same with an average area of permanent pasture which produced 17 cwts per acre.

The prospect of needing to make more hay meant that the process also needed to be speeded up a bit especially the mowing operation and I began to entertain the thought of a second-hand tractor to replace three of five cart-horses we were keeping, and perhaps doing a bit of contract mowing etc. The difficulty was that pneumatic tyres on tractors and farm machinery had not yet arrived, and there were only two fields on the farm that a steel wheeled tractor could get to without going on a public highway. However, enquiries revealed that Ballards, the agricultural machinery firm of Abingdon had a second-hand International 10/20 on steel wheels, the rear ones equipped with rubber strakes cut from the solid rubber tyre rims of wheels from an old steam propelled road wagon. A meeting with Teddy Ballard who owned the business which, before moving to Ock Street, was situated on the edge of the old cattle market (now a shopping precinct) revealed that although the outfit appeared a bit ungainly, it would travel along the highway without causing too much damage and solve our problem. The engine was started up on petrol and when warm switched over to T.V.O. (Tractor Vaporising Oil). Adjourning to the Queens Hotel run then by Henry Wilkes in a very efficient manner and operating a "Men Only" bar, the nitty gritty of a possible deal ensued. Teddy Ballard wanted £50 for the tractor, father bid him £40 — three or four half cans of beer later, father raised his offer to £45, whereupon Teddy said he would accept the £45 offered, but he would need the balance of £5 in 12 months during which time he would honour a reasonable guarantee to keep it serviceable. A slap of the hand and another half-can each clinched the deal. The old tractor behaved perfectly for a lot longer than 12 months. Teddy duly collected his £5 and replaced the 10/20 with a brand new W14 of the same make complete with pneumatic tyres three years later.

The next rung on the expansion ladder was the purchase of half a dozen freshly calved heifers. These were selected from a bunch of fourteen tied up in a cow shed at Radley for us to see by a well known and respected cattle dealer named Dick Greening. He was a charming character, always well turned out and seldom without a flower in his button-hole. Father having explained to Dick that he wanted me to have the experience of picking them out, they left me alone in the shed telling me to take as much time as I needed, and the price for the six of my choice would be £25 apiece. Meantime, they would be indoors having a drop of gin and I was to join them when I had made my decision. Having made my choice, Dick remarked that he would have picked 5 of them but thought the sixth would put a lot more on her back than in the bucket. "Time alone will prove you right or wrong my lad but there is no better way of learning than paying for your mistakes, so you take her home and find out." So I did — and he was right!

It was about this time, when things seem to be developing nicely that a few clouds began to appear. First of all my dear mother had not been feeling well for quite a while and our family doctor, Dr J.B.Pulling decided she needed to go into the Radcliffe Infirmary in Woodstock Road, Oxford for observation, which sadly resulted in a diagnosis of diabetes, consigning the poor dear not only to a fairly strict diet, which she coped with admirably but the reliance on two injections of insulin daily, the effects of which required close monitoring by simple but regular tests at home. O how she hated giving herself the injections, which sometimes were exceptionally difficult to achieve as she suffered a great deal from rheumatism as well. There were times when she pleaded with father or me to do it for her, but somehow neither of us had the courage. I did try a few times but gave up saying "I can't bear hurting you but I know I could do it to someone else."

Dr Pulling who practised at Faringdon could well be described as a "workaholic"; he was absolutely dedicated to his calling. As transport into Faringdon from neighbouring villages was difficult he seemed to carry out most of his work by visiting his patients at their homes, attending various areas on certain days and evenings of the week, and one often heard someone say "It's Doctor's night to-night". Medicines were dispensed at his home in Church Street by the local chemist Billie Cook and delivered to patients by the postman. I know it is true that on some occasions half-a-crown had been discovered on the mantelpiece of more than one cottage following a visit from this quiet, kind-hearted man. I remember mother telling me that he admitted not going to bed on some nights, when he needed to finish writing important letters; so in order that he would not drop off into a long heavy sleep, he would wake himself up by having tied his bootlaces up very tightly which after a short while caused pins and needles in his feet. Very sadly he lost his wife during the birth of their first born and this almost fanatical devotion to his work was the only way in which he could find comfort in his grief. His baby son survived and was virtually brought up by a delightful lady named Nanny Watford who also kept house for Dr Pulling until he died. She was well known in Faringdon for her kindness and devotion not only to Doctor and his son but to the large number of patients with whom she came into contact. The only real relaxation Doctor had, as far as I know, was a fortnights sailing every summer which he looked forward to and enjoyed enormously, a fact which was underlined by the large number of pictures of sailing vessels which decorated the walls of his home. It is possible too that he gained a certain amount of relaxation from tobacco, as he once told me he looked forward to a cigarette between calls on his rounds, smoking a 'Player' on the longer journeys, while a 'Woodbine' sufficed on the shorter ones. He was also very fond of stopping on Saturday afternoons for a short while to watch a local football match, where he would position himself on a deserted area of the touchline, so that he could retrieve the ball when it was kicked out of play. This seemed to please him a great deal and I have often seen him do it when playing for Uffington or Stanford in the Vale. In 1936, he became the first President of the Faringdon Thursday Football Club, which later organised the Faringdon Hospital Cup Competition. While playing for Stanford we won this cup beating Watchfield in the final at Tucker Park, and I had the honour of lining up with the team to be presented with small individual replicas by the kind doctor himself. Stanford had a good season in 1937/8 winning the Swindon & District League, the Dr Elliott Cup and only just failed to win the coveted Swindon Advertiser Cup on the County Ground being beaten 2 — 1 by Wroughton, "The Football Pink" reporting that we were a bit unlucky!

During the first few months of Mother's struggle to come to terms with being a diabetic she was helped a great deal by Lena Ackrill, an unmarried daughter of Henry Ackrill who I mentioned earlier. Lena being middle-aged and still living in the cottage where she was born, was a tower of strength, looking after us and helping Mother in so many ways.

About this time another cloud had started to appear, this time on the farming front. It was known plainly as Contagious Abortion in cattle and caused in-calf cows and heifers to abort their calves around about the middle of their pregnancy. At that time there was nothing that could be done to prevent this happening other than perhaps a long term attempt to make one's herd self-contained but as the foetus, afterbirth and subsequent discharges were likely to carry infection, spreading by birds and foxes was a distinct possibility this approach was far from being certain to eliminate the problem. There were a few proprietary drenches which were supposed to help but which were in fact a waste of money. As we had been free from the disease before buying in cattle to increase our herd size, we concluded the "bug" came with them. It started by affecting the younger cows, mostly the second calvers, playing havoc with our calving pattern and production targets, finally becoming evident for the first time in a bunch of first-calf heifers, which were on rented grass at Dick Brewers Spanswick Farm, Letcombe Bassett.

In consultation with our local Vet and the Ministry of Agriculture Veterinary Department who had already indicated that in time the trouble would blow itself out and the herd would eventually build up its own immunity, it was thought best not to bring any of these heifers which were due to start calving in October, back home to Church Farm, in order not to add further fuel to the fire but we hoped would also help to shorten the period of recovery.

So in the summer of 1938 we swallowed the bitter pill and sold the twenty in-calf heifers, one of which had already aborted, to Claude Ryman, a cattle dealer from Didcot for £18 apiece, about two thirds of their value, with full-time calves by their side. Claude Ryman called on us regularly. We had many deals with him and he would buy or sell almost anything in the farming line. He told us later only one of the heifers carried her calf to full time and the rest he allowed the milk, which was usually a very small amount, to dry up, put a bit of meat on them and sold them in the late autumn. Of course he also told us they lost money!! We liked dealing with him as he rarely criticised what he was being shown before making an offer which he thought was not only fair but open to reasonable negotiation, and he always paid "on the nail". Sadly, he died in middle-age from a heart attack. I remember his father who was also a farmer and dealer with a special interest in horses. The first time I met him he had come to Church Farm to look at a cart-horse, which unfortunately had gone with a trace-horse to Field Farm, Longcot to fetch a load of thatching straw. Consequently father sent me with Mr Ryman to show him where we were likely to see the horse, so I climbed up beside him in his old Ford tourer and the first thing I noticed was that he had a knife and fork sticking out of his breast pocket. Upon inquiring about them he told me he always carried them with him in case anyone offered him a meal. He was quite a character and spoke to his car, as one would while driving a horse saying "come here" if turning to the left and "get over" when turning to the right and "steady", — "Woa" when he was going to stop. We met the load of straw on what is now known as the A420; our carter Bill Mace suggested he went on a couple of hundred yards or so and turned right, down the Longcot road, take the horse he had come to see out of the shafts, trot him up the road or test his wind by holding him alongside the wagon, taking hold of his bridle and threatening him quickly with his stick (most dealers carried a stick of some sort) to frighten him. This was done two or three times and if the horse grunted he was suspected of being "wrong in the wind", a horse dealer's expression of indicating some sort of breathing difficulty.

Mr Ryman said he hadn't time to wait for that, so the matter was proceeded with on the spot and was concluded by giving the carter a shilling for his trouble and me sixpence upon returning to Church Farm, where a deal with father was struck, which if memory serves me right, was £25.

It was now 1938 and Herr Adolf Hitler was causing all sorts of trouble in Europe, annexing territories belonging to other small nations by sheer force of arms. A situation which many were already convinced was certain to lead to a second World War, and for which the rest of the world it has to be said was ill-prepared. So it was that various defence schemes were hastily set up by the Government to deal with air attacks and the possible use of poison gas on this old island of ours. Hitler's Italian partner in crime Benito Mussolini was also bawling his head off at military parades in Rome, following his hostile seizure of Abbisinya. He couldn't have picked a much weaker country to invade, but to listen to his ravings one would have thought the United States of America was next on his list.

I think the first uneasy feeling I experienced concerning Hitler's intentions was while being shown round London's first international airport at Croydon not long after Amy Johnson's epic Australian flight. It was a conducted tour and I well remember our guide saying the only part of the airport which he was unable to show us was that occupied by the German 'Lufthansa' Airline Company, mentioning that it was possible they were not particularly anxious for the general public to observe how quickly their passenger aircraft could be converted for military purposes.

While we were looking round, ground visibility was quite good but the cloud ceiling was very low, concealing the arrival of what we were informed was the first of the daily flights from Paris. As it passed unseen a few hundred feet overhead we were told that it was the Handley Page Heracles, piloted by British Imperial Airways No 1 pilot Captain Jones (I think it was); he had flown more hours than anyone else and was recognisable by the beard he wore! He was, we were told, in wireless communication with the control tower and was being "talked down". It was quite a sight to see this massive ungainly looking four engine bi-plane drop out of the low cloud, make a perfect landing and taxi on to the tarmac to disgorge its relatively small number of passengers, each one of whom I looked upon as an adventurer, rather than someone who wished to get from Paris to London in the quickest possible way.

During the autumn of 1937 I was asked with a few others of various ages, by the local policeman if I would like to attend a meeting at Faringdon Police Station one evening for the purpose of joining a defence organisation to be known as the Observer Corps whose duty in the event of war being imminent would be to man a look-out point near Uffington, continuously 24 hours a day, for the purpose of reporting, all aircraft that could be seen or heard in the area by direct telephone line to a Group Headquarters to be set up at Oxford where all movements would be plotted on a table map and passed on to R.A.F. Fighter Command, at Uxbridge.

After the meeting, about a dozen of us volunteered to join the Observer Corps that evening. Following Anti-Gas training and a concentrated course on aircraft recognition mainly by the use of silhouettes we were all sworn-in and issued with

Warrant Cards as Special Constables of Berkshire to enable us to carry out our duties without undue hindrance. The look-out post was set up in a field to the south of the smallholding in Uffington, farmed by the Parrott Bros, George and Fred. It was approached by a footpath and as special constables we had the right to challenge any unknown characters in its vicinity. The Post was identified as X2 and manned at all times by a team of two, any such challenge as "halt who goes there" was backed up one of us holding an army rifle with a couple of "blanks" in the magazine. In fact the post was issued later with a few round of "live" ammunition but I don't recall the trigger ever being pulled in anger.

On one occasion while on night duty with "Longie" (Mr Long) the head gardener at Kingston Lisle Park, a lone German Heinkl III came hedge-hopping westwards at an altitude of about 150 feet and passed within a hundred yards of the Post. We knew it was approaching as we were able to hear Post XI (located at the top of Chain Hill just south of Wantage) reporting its progress towards us. "Longie" who had served as sergeant in the First World War with distinction took immediate command and promptly told me to follow his example by donning one of the two steel helmets provided. As it was a bright moonlight night I suggested "Jerry" might present a possible moving target for a few pot shots as he flew by us, but my dreams of being the first and probably the only Observer ever to shoot down a hostile aircraft were quickly dashed by "Longie" telling me not to be so foolish, adding that if I did manage to put a bullet through the fuselage somewhere, "Jerry" could well take umbrage, and come looking for us for the purpose of getting his own back. A chilling thought on reflection. However, I did reap some benefit from the incident, as the next time "Longie" and I were on duty together he said he hoped he hadn't upset me by disciplining me and gave me a beautiful home-grown cucumber as a peace offering. A rare treat it was too as they were in short supply during the war years.

Another incident happened in the early days of the war before the fall of France when there was a limited amount of activity by the Luftwaffe during the hours of darkness especially if the weather was stormy. The layout of our observation post at the time consisted of a small wooden hut roughly 12 ft square, equipped with a single bunk, an old armchair, a coal burning stove for heating, basic cooking and boiling a kettle. The concrete pad on which the tripod holding the instrument was situated about 20 yards away. This was used to plot aircraft by sight or sound causing a pointer to move across a sectioned map of our area, enabling the second member of the team on duty to report the aircraft's movement by telephone. To provide some shelter from the elements the instrument area was surrounded by a few sheets of galvanised iron with a staggered entrance, the whole of which being referred to by John Betjeman as "the Urinal" when he served as a member of X2 before joining the Ministry of Information. It was on a very foggy night that John and I were on duty and as there was no flying activity we had retreated to the warmth of the hut. While I admit to taking "forty winks" on the bunk, John, who was reading a book, had the headphones on and said he would wake me should we be required by Centre, located on the top floor of Oxford Post Office. At about 3am John shouted that Centre were requesting us to check westwards as Bristol were reporting the sound of an aircraft believed to be hostile. He dived out into the darkness through the hut door swathed in about 25 yards of telephone cable which by the time he had got half way to the "urinal" had tightened around his legs and brought him to the ground, shouting to me through the fog, "My God, my God, we have lost the war, we've lost the war". It was at about this time during similar weather conditions that I amused myself during a period of night duty by composing the following:—

ODE TO X2

P is for PACKER, the head of our Post
Of the worry entailed, He gets the most.

He hands us the "dough", decorates the 'ut walls,
With duties and notices, and is "BILL" to us all.

There is one more PACKER, and ERNIE's his name,
He's a relation of BILL's, but not quite the same.

B is for BASSETT, at darts he is tough,
When a "double" he wants, one dart is enough.

B is for BETJEMAN, of "charring" he's fond,
The floor while he's at it, is like a "girt" pond.

F is for FIELD, who makes quite a "splut",
He just gets on his knees, and scrubbs out the 'ut.

G is for GRANGER, At Shrivenham he settles,
Now the petrol is rationed, he "burrs" on the pedals.

I is for ILES, Just Bob for short,
His usual drink, one straight Quart.

L is for LONG, At plotting he's hot,
If there's anything doing, He's right on the spot.

L is for LIDDIARD, who's hair is not curley,
He's usually late, and is NEVER early.

L also for LEAHY, t'was a bit of a lark,
Was late for night duty, got lost in the dark.

R is fro RICHENS, and ROBERTS as well,
If ever they're parted, there's sure to be Hell.

W is for WILLIS, who is usually FLAT OUT,
Has a liking for Wine, of which there's no doubt.

Now W for WENTWORTH, A butcher worth trying,
Is never happier, than when he is frying.

W is for WHITEHORN, and Willie too,
Call him one or t'other, either will do.

Now for the WEAVERS, our one family,
George, Edmund and Tom, a total of three.

Last but not least, is WESTCOTT our friend,
A message to whom, we should like to send.

Here is the same, it is quite short and sweet,
Go careful my lad, when you sit on the seat.

I think you'll agree, we're a reet motley crew,
But ther's no post on earth, LIKE GOOD OLD X2.

ONE OF 'EM

With apologies to Nobody.

Each duty consisted of 8 hours, the first starting at 6am until 2pm, followed by 2pm to 10pm and the night duty from 10pm to 6am. I was in fact afforded the consideration of not being asked to do the 6am to 2pm stint to enable me to take part in early morning milking at Church Farm.

The first year of our training consisted of occasional manning exercises and visits to local aerodromes which often included being taken up for short trips in various types of training aircraft, such as a Miles Magister, a small low-winged monoplane which was my favourite. It seemed to provide all the joys of being airborne coupled with an almost unhindered view of the ground below. We also flew in Oxfords and Ansons from R.A.F. Watchfield. Later during wartime some of us were taken on a round trip training flight in a Wellington bomber from R.A.F. Harwell to Lincolnshire. Being shown the basis of the operation and allowed to sit in the desperately lonely rear gun turret for a while, brought home the stark reality of it all and the tremendous courage being displayed by everyone involved in operational duties of all kinds.

Prior to these trips, I had only flown once. It was when Sir Alan Cobham visited the locality in the mid 1930's with his Flying Circus air display, flying from the large field between Kitemore House and Shellingford cross-roads. I can't remember the make of the bi-plane I went up in but I do remember it had three separate in line open cockpits, and the five minute trips cost five shillings (25p). If a trip rather more free from the sound and buffeting of the wind was considered more desirable, then the cabin of a six-seater De Haviland was on offer at seven shillings and sixpence (37½p) for a similar length trip. It was quite exciting, and a topic of conversation locally for quite a while afterwards. The pilots were kept very busy during the three afternoons and evenings flights were available; after which the Circus moved on elsewhere.

My very first flight as a member of the Observer Corps was for a very good reason particularly memorable to me. It was from R.A.F. Abingdon in a two seater open cockpit Gloster Gladiator. The pilot, having inquired where I lived, circled Baulking a few times giving me the unique opportunity to see Church Farm from the air for the first time, as well as wave to mother and father on the lawn! On returning to Abingdon and circling the aerodrome a few times waiting for permission to land I recognised Dunmore Farm which was only a field away from the aerodrome perimeter. It was owned and farmed by the Benson family. Having lost their father who was also a well known heavy horse dealer, the business was carried on by the eldest son Jack and I had been there a few times as a young lad with my father to look at possible purchases. It also came to mind that, as best man to a friend of mine "Podge" Hawes, a farmer's son from Horton-cum-Studley, who was assistant manager at Barclays Bank at their branch in Faringdon and that I should be attending the wedding reception there in three weeks time to celebrate his marriage to one of the Benson girls, Betsy, whom he met when they were both working at Barclays in Banbury. WHAT I DIDN'T KNOW was that I was destined at the age of 25 to meet one of Betsy's younger sisters Ann and that in fourteen month's time I should be attending a similar celebration there in the much more important role of bridegroom.

SEVEN
Taking The Plunge

Ann and I were in fact engaged early in July 1939 and married in Shippon Church, a stones' throw from Abingdon R.A.F. Station on November 16th 1939.

Since then by the Grace of God, we have been privileged to spend fifty-eight years together, sharing many joys and sorrows too. We have been blessed with a son and daughter who have both found loving and caring marriage partners resulting in five lovely grandchildren and we are proud of them all.

Being wartime, our wedding was a quiet family affair. Food, clothes and petrol were strictly rationed, obtainable only by coupons. Apart from petrol for travelling necessary in connection with one's business and for use in stationary and portable engines for driving barn machinery milking machines etc., a very small amount was granted for personal use such as shopping if one lived in an isolated area or a village without shopping facilities.

There were very, very few diesel powered farm tractors pre-war, the only one I can remember was a Lanz, which belonged to H. Baylis & Son of Hatford and powered their threshing tackle. Its engine was started by unscrewing a cartridge holder on the top of the cylinder head inserting a roll of ignited touch-paper screwing it back in position and quickly turning the engine over a few times to build up sufficient compression in the cylinder to enable the fuel from the injector to combust. The winding handle fitted into a heavy cast iron flywheel which, when in motion, added impetus to the operation but it could be a bit exhausting on a cold frosty morning when the oil was a bit thick, especially if the chap on the handle happened to be suffering from the effects of a hangover.

Most tractors ran on Tractor Vaporising Oil, (paraffin in other words) and it was necessary to prime their engines with a small amount of petrol to enable the vaporising unit to become warm enough to vaporise the lower octane fuel when it was introduced.

The change over from petrol to T.V.O. was achieved simply by means of a small three-way brass tap. Pre-war the small amount of T.V.O. remaining in the carburettor float chamber, about an egg-cupful, was drained on to the ground via the drain plug but someone calculated that if all these egg-cupfuls were collected in a cocoa tin, or similar, suspended beneath the carburettor with a piece of wire and they were tipped straight back into the main supply tank it would save hundreds of tons of crude oil going to waste and in a small way contribute towards the war effort by helping to alleviate the tremendous strain on the merchant tanker tonnage which despite sailing in convoys protected by the Royal Navy suffered considerable losses at the hands of the German U Boats.

Despite various difficulties, Ann and I had somehow managed to save up enough coupons to purchase a wedding outfit each and sufficient petrol to get us to and from the Livermead Cliff Hotel, Torquay, for a long weekend. Due to it being mid-November, black-out time was about 4.30.p.m. and it was difficult and hazardous to drive too many miles by the light of three narrow slits in two headlight masks, so we decided to get as far as "The George" at Amesbury and complete the rest of the journey to Torquay the next day. Having dropped father and mother off at Church Farm after our wedding I remember father pressing a £1 note into my hand saying "Have a bottle of wine on us this evening and good luck to you both". The total cost of our overnight stay was £1-13s-3d, covering apartments, Dinner for two, a couple of drinks, a bottle of wine (3/6d!) breakfasts and a garage fee of 1/—.

Being close to the Army & R.A.F.Camps on Salisbury Plain,

the Hotel was full of Service Personnel, presenting an atmosphere deeply indicative of the apprehensive mood into which everyone had been so suddenly cast. Not a particularly pleasing prelude to the first night of ones' honeymoon.

EIGHT
World War II 1939-45

I have vivid recollections of the declaration of war with Germany. It was delivered some two months earlier in a speech made to the nation, over the radio, by the Prime Minister, Neville Chamberlain on the morning of Sunday 3rd September 1939. I was on Observer Corps duty with an estate agent named Norman Granger who had served in the First World War and was on the staff of Hobbs & Chambers at Faringdon. It so happened that the battery on the somewhat old wireless set, which someone had kindly presented to X2, was getting a bit flat, so we kept the volume down to conserve as much power as possible. Crouching over the set together it seemed the whole world came to a standstill as the words "This is London" fell from the lips of John Snagg the B.B.C.'s senior announcer. Somehow his voice had the immediate effect of galvanising those listening into a state of dramatic expectancy, in readiness for the importance and total reliability of what they were about to hear.

We were not to know, during the following five years those words were to be repeated on many historic occasions as preludes to Christmas messages by King George V1 from Sandringham and rallying calls from the great old bulldog himself, Sir Winston Churchill when he became Prime Minister of the Coalition government formed to lead us through those dark days. He quickly told us at the outset he could promise us nothing but blood, toil, tears and sweat, we would fight to the bitter end and would NEVER EVER surrender. What a rousing patriotic speech it was.

Despite the war we spent three happy honeymoon days at the Livermead Cliff Hotel, Torquay, which I see from the letter confirming our booking was not too costly and I quote:—

One double bedded room, facing the sea for Nov. 17th, 18th, & 19th, at the inclusive rate of 15/— (75p) per person per day.

Lunches (HOT) 3/6 (17½p)

Table d'Hote Dinners, including coffee served at 7.p.m. 6/— (30p.)!!

On our return to Church Farm we spent a short while with mother and father until they were able to complete their move to Sherbourne Cottage, Uffington, a property comprising a three bedroom house, built just after the First World War, and a fair sized lawn and garden. It had been offered to them for £900 but the site was copyhold in favour of a very long lease granted by the Craven Estate, meaning that any tenant would find it difficult to substantiate a tenure, having nothing to show except the copy of a Court Roll. Father was therefore advised not to buy the property but did manage to rent it for £50 per annum, plus rates.

My solicitor friend Sam Wilkinson tells me this particular legal state of affairs had its origin in the time when it was the privilege of the owner of the manorial rights, the Lord of the Manor, to grant a favour to someone, such as an estate or farmworker, permitting him the use of a small plot of land for his own benefit, for which he agreed to pay what was known as a Quit Rent, enabling the Lord of the Manor to regain immediate possession at anytime, should he so wish. Apparently, down the years, the increasing fragmentation of large estates led to difficulties within the transfer of legal ownership's, so a move was made in 1922 to phase the problem out, but it was not until the 31st October 1950 that it was finally completed.

During the early weeks of the war the A.R.P. (Air Raid Precautions) organisation, which had been formed early in 1938 as a precautionary measure prior to the Munich crisis, was mobilised in strength to do its best to alleviate the consequences of probable air raids, not only on our cities and towns, but other likely targets such as railways and the many airfields which were being hastily constructed in rural areas.

The A.R.P. was staffed by volunteers who after short training courses were appointed Air Raid Wardens. It was their task once the air raid siren had sounded, to get people to put on their gas masks and shepherd them to the nearest air-raid shelter. It was then their duty to patrol the area to which they had been allocated for the purpose of reporting any injuries or damage caused by high explosives, incendiary devices or poison gas. Mercifully, the latter was never used.

In order to test the readiness of the organisation, a full scale night exercise was arranged to respond to a local mock enemy air attack which was supposed to have been carried out on Buscot Park, the stately home of socialist peer, the late Lord Faringdon. Summoned by the wailing of sirens everyone concerned converged on the area which was strewn with bodies, some beyond help but scores of others bearing labels indicating what injuries they were supposed to be suffering while lying in the darkness of the gardens and precincts of the big house, waiting to be rescued and treated.

One such "victim", having been found on the front lawn by the dim light from a masked torch which also provided enough light to read the attached label, necessitated the finder shouting for a doctor to deal with extensive facial injuries and a broken nose. This brought our well loved local Doctor Pulling, clad in gas mask and steel helmet to the victim's side; unfortunately as he knelt closely over him to examine the damage, his steel helmet being much too large for a man of his slight build and stature fell off, the steel rim striking the poor "casualty" on the bridge of his nose, quickly turning fantasy into painful reality.

The first few months of the war produced very little action with the Allies ranged behind the supposedly inpenitarable "Maginot Line" built by the French to protect their eastern frontier for all time following the First World War and the

Germans massed behind their own similarly fortified defences named the Zeigfried Line. This period became known as the "Phoney War". In fact the 1939/40 winter was a very cold one with a great deal of snow and frost, some of which produced a phenomenon brought about by the freezing of rain as it fell, causing thick coatings of ice to build up on telephone and electricity cables, boughs of trees etc., many of which broke under the extra weight. It also had the effect of making it impossible to drive a car or lorry or even to stand on one's feet due to formation of black ice on roads and paths etc.

The German propaganda machine under the direction of Goebels, had been boasting on their short wave wireless programme that Hitler had a secret weapon up his sleeve and it must be said there were even one or two who were daft enough to believe that this was it! Most of this rubbish came from a regular daily war news service broadcast for our consumption by a traitor named Joyce, who became known as Lord Haw-Haw, mainly because of his "hawty" toned voice as he introduced the transmissions by saying "This is GARMANY calling — GARMANY calling" on a wave-band of whatever it was which was usually followed by incoherent statements, mostly false and of course designed to undermine morale on the home front. Often reception was impossible due to our people "jamming" their efforts completely.

My father joined the A.R.P. about the same time as I joined the R.O.C. early in 1938 just after Hitler had marched into the Ruhr and was looking towards Czechoslovakia.

To boost recruiting for our home defence services a huge rally of existing members was staged one Sunday in Hyde Park. The event culminating in a march past Buckingham Palace and up The Mall to Trafalgar Square, each county's contingent of members being preceded by a banner bearing the appropriate name. As our part of the world was then in Royal Berkshire and not Oxfordshire as it is now; (that it had to be changed I have yet to reason why — I still like to think I am Berkshire bred and born) father marched a Berkshire man. I remember him saying it was quite an experience and recalling the remark of one cockney to another as they saw the Berkshire banner approaching "Here come the Beer, Bacon & Biscuits Brigade". indicating that they were well aware Reading was the home of Simmonds Brewery, Venners Bacon Factory and Huntly & Palmers.

Someone once told me a story concerning the latter. Apparently a few years ago the method of biscuit-making at Huntly & Palmers was considerably updated by scrapping most of the old machinery, installing modern equipment complete with automatically controlled ovens. Unfortunately, when production recommenced it was difficult to perfect the cooking; the biscuits being either over cooked or under cooked. An attempt to solve the problem resulted in sending for the recently retired baker/cook, who said when confronted with the paraphernalia of modern technology "I don't understand all this, I could always tell when the biscuits were ready by just sniffing".

With the winter months over the "Phoney War" came to an abrupt end with a German blitzkrieg resulting in the rapid and overwhelming defeat of the low Countries and France, leaving the British Expeditionary Force no alternative but to retreat towards the beaches around Dunkirk from where mercifully most of them were evacuated by a fleet of every kind of ship or boat that could possibly float.

Immediately following this, it was generally considered that the enemy would attempt to gain air superiority by knocking out our small R.A.F. Fighter Force, then stage a sea-borne attack on the South and East coasts, at the same time dropping large numbers of air-borne troops inland. So it was that on 14th May 1940 Mr Anthony Eden, then Secretary of State for War announced by a broadcast to the Nation that a Force to be known as The Local Defence Volunteers was to be organised in Britain. Their role would be to combat any landings in this country by enemy paratroops.

Any able-bodied male of 17 years or over would be eligible

to join and volunteers were directed to report to local police stations for instructions. Arms would be immediately available and uniforms in the form of khaki denim overalls and service caps would be issued in due course; in the meantime special L.D.V. armbands would be worn.

On hearing this, my father, having lived in the countryside most of his life, considered he would be of more use in this kind of role than in the A.R.P. so he transferred immediately to the L.D.V. and was put in charge of the hastily assembled Uffington Platoon.

One of his first tasks was to organise a Guard consisting of one N.C.O. and four men. Their job was to patrol the area on the downs around White Horse Hill, keeping a sharp look out for anything unusual. They went on duty at 10.p.m. until 6.a.m. and their quarters during this period consisted of an old iron-wheeled shepherds hut. These "guard duties" were later shared by personnel from other platoons in the area and the L.D.V. eventually became known as the Berkshire Home Guard when, in accordance with War Office policy, it became attached to the Royal Berkshire Regiment and wore the County Regiment Badge.

Another task of platoons was to construct road blocks at the approaches to all villages. These consisted of all sorts of horse drawn farm implements such as side rakes, hayrakes, cambridge rolls, those of the "Tree & Wheel" variety — consisting of a fir or larch tree trunk measuring not less than 12 inches at base. It was hinged to a stake on one side of the road with a wheel at the other end so that it could be quickly rolled into position across the road and secured to another stake on the opposite side. These somewhat skimpy barricades were reinforced with coils of barbed wire, while slit trenches were dug nearby so that cross-fire could be brought to bear on any enemy troops or vehicles attempting to pass. John Florey of Duxford told me that one such obstacle was situated outside the farmhouse of Wilfred Hawkens at Hinton Waldrist and he used to roll it into position before he went to bed every night, and open it up again first thing in the morning.

I remember father having what he thought was quite a bright idea by asking Bert Ayres, who lived in the last house going out of Uffington on the White Horse Hill road, if he could sand-bag a very small attic window which faced southwards [from his house] commanding an excellent view of about half a mile of straight road leading into the village from White Horse Hill. "Just the place to pick the buggers off if they land somewhere up there" he said. What with I have no idea as at this time the Wantage Battalion, made up of fourteen hundred men who had volunteered in the first few days, were virtually unarmed. Three Hotchkiss machine guns had been issued, but these were of little value in view of the lack of ammunition to go in them and the complete lack of trained crews to use them. In addition there were 120 shot guns loaned by farmers and local people, in response to a Government appeal, for which a minute quantity of ammunition which became known as "12 — bore Lethal" was issued. It consisted of a 12 bore cartridge case in which the lead pellets had been replaced by a single lead bullet about the size of a gob-stopper. The whole thing was reported to be as deadly as it was unreliable and prone at times to split barrels of shot-guns from end to end. These few weapons, together with about half a dozen privately owned revolvers, were the sum total of the Battalions armament. Those who had no fire power relied on three dozen Mills hand grenades and [some] home-made "Molotoff Cocktails" (a mixture of tar and petrol corked up in an old bottle, said to be most effective against tanks). There was a great deal of rumour that more suitable supplies of armaments were on their way but alas this was not borne out by facts, one of which I relate. Father told me that his Company Commander, Capt Tom Allen-Stevens of Wicklesham Lodge, Faringdon, had phoned to say he was making a weapons delivery in the evening. Father immediately expected a fairly large consignment of some rather more sophisticated equipment such as sub-machine guns, complete with ammunition etc. so he cleared out his garage at

Sherborne Cottage and asked me if I would call in on my way home from Observer Corps duty to help him unload. Having arrived on my bike I was followed into the drive by Capt. Allen-Stevens in his Ford V.8 two toned brown Estate. Father emerged from the house and excitedly waved the Captain to reverse up to the already open garage. Having done that, Captain Tom unlocked the rear doors of his Estate to reveal three dozen lengths of 3/4 inch galvanised piping with spikes riveted in one end, all of which had been fashioned that morning by the local blacksmith Mr Ballard of Coxwell Street, Faringdon. To add to father's disappointment the Captain explained that he could only leave a dozen as the others were for Shrivenham and Ashbury platoons. Father's inquiry as to what he was supposed to do with them, was met by "I don't really know Cyril, but if you can get round behind, they won't half hurt".

Although the task of unloading was far from being laborious I seem to recall it was followed by the consumption of one or two Gin & Gingers before the Captain left for Ashbury. As I pedalled my way home to Baulking I could not help wondering how under these circumstances a miracle would be needed to save ourselves from the well trained and resourceful German Paratroops armed, no doubt, to the teeth with the most modern and deadly weapons. I remember joking to Ann that we had better take a couple of four-grained prongs with us when we went upstairs to bed that night.

It is now common knowledge that at this time and for some time afterwards there were only two Field Army Divisions in the British Isles, one British and one Canadian and the number of tanks could virtually be counted on one hand, so perhaps it was indeed little short of a miracle that the expected never happened, due no doubt to the victory gained by those courageous members of the R.A.F. who fought like terriers as the battle raged in the skies above us and of whom Sir Winston Churchill said later — "Never has so much been owed by so many to so few".

There was a nation-wide order to muster one Saturday evening in September but this turned out to be a false alarm and was followed by a "stand down" order the following afternoon.

There were of course a few relative "incidents" locally. Father received a message from the police late one afternoon to say that paratroops had been seen dropping in the vicinity of Knighton Bushes, close to White Horse Hill. By the use of bicycle and telephone he managed to gather a few of his platoon together, the first to arrive at Sherborne Cottage being his Section Leader Joe Bailey. Father handed him a shot gun and a few rounds of "12 Bore Lethal" and suggested he should take up a particular observation point, where he would be joined later by other members of his section as soon as possible. Off he went on his bike, only to return to father a few minutes later to inquire "If he saw anyone, was he to shoot to kill or just hurt 'em?" It was a mercy he didn't see anyone at all as the "enemy" turned out to be a platoon of Canadian Paratroops on a training exercise.

Another incident illustrating everyone's expectancy of an invasion occurred when I was returning home through Uffington in the car after an eight hour shift at the Observer Post. It was in the black-out on a very dark night at about 10.15.p.m. I was following close behind another car which had just emerged from the Fox & Hounds. About a hundred yards from what is now the Thomas Hughes Memorial Hall we were stopped in our tracks by a blinding flash and what seemed like an explosion. After a minute or so the occupants of the car in front of me got out and by the dim light from my masked headlights I could see they were two army officers (who I learned later were on an O.E.R. course at the Military College at Shrivenham). One of them had a torch so the three of us, suffering it must be said with a certain amount of the "shakes", advanced up the road on foot to where we all expected to see some sort of bomb crater. However what had actually happened was that a huge branch from an old elm tree had broken off and fallen across the road bringing down the electricity supply lines, which had fallen into the brook by the side of the road, causing the flash and explosive sound as it did so, as well as plunging the whole village into darkness. This unfortunate incident was the cause of a great many recitations in the four pubs and two village shops for a day or two, including one from old Harry Bailey, a retired ganger on the Great Western Railway , who lived with his dear wife Belinda, in a spotlessly kept cottage and garden within a few yards of the disaster. Harry was heard to say in the Bakers Arms the following evening that they had just gone upstairs to bed and that mother (as he called her) had just got herself between the sheets when all of a sudden there was this hell of a bang and all the lights went out leaving him groping about in the dark, shouting "Get thee-self up mother, the buggers be come".

There were of course many stories relating to Dad's Army, as the Home Guard later became known, one of which I must relate. It was told to me by the late Jack Roberts of Ashbury but I must admit to having some doubt as to its authenticity.

It was said to have taken place in the winter of 1941 when the Home Guard had become better organised and reasonably well equipped. A youthful agricultural worker and member of the Ashbury platoon was patrolling the outskirts of the village, prior to meeting up with a colleague at a pre-arranged venue. It was in the early hours of a clear frosty winters night that he found himself in a field adjoining the bottom of his family's cottage garden. That evening he had an altercation with his father who was a shepherd, and had lived in the old thatched farm cottage all his life, concerning the somewhat crude toilet facilities with which they had to cope, in comparison with those available to town dwellers and why couldn't something be done about it? It was therefore perhaps understandable that the sight of the ivy-clad privy a few yards away was too much for him. Impulsively he detached a hand grenade from his belt, removed the pin and tossed it into the ivy on the privy roof and lay flat on the ground. When the sound of the explosion subsided all the dogs in the village started barking and the owls hooting, and there in the moonlight, sitting on what was left of the seat with his face blackened and night shirt in ribbons was his ol' man. Seized with spontaneous remorse, the young man ran to him, threw his arms around him shouting "whatever happened Dad?" — "I don't know" replied the old chap — "I think it must have been that tapioca pudding your mother gave us all for tea."

As Churchill promised, the first three years of war were dark indeed but a tremendous spirit of comradeship developed, thousands of children from the cities and large towns were evacuated into the comparative safety of the countryside and most of them learned for the first time that milk came from a cow and not out of a bottle.

All sorts of schemes were organised to raise money and produce more home grown food. These were promoted with slogans such as "Wings for Victory" & "Dig for Victory" etc. Scrap metal was a valuable contribution to the production of armaments and was collected regularly from farms and villages by members of the National Farmers Union who also organised sales and events in aid of the British Red Cross Fund; oOne of those being the presentation of a variety performance by the famous dance band leader Jack Hylton at the New Theatre, Oxford, now the Apollo. Among those taking part were Flanagan & Allen, "Stinker" Murdock, Adelaide Hall, Florence Desmond, Arthur Askey, Dr Malcolm Sargent, conducting the London Philharmonic Orchestra.

Those of us living in the countryside developed a tremendous admiration of the city dwellers who suffered the might of the Luftwaffe almost night after night. To illustrate the point I remember Air Marshall Sir Phillip Joubert addressing a large gathering in Faringdon Market Place to open a "Wings for Victory" week for the purpose of encouraging everyone to invest as much money as they possibly could in National Savings to support the war effort — in this case, to help provide the aircraft we desperately needed to keep Hitler at bay.

He said in his speech that he had been summoned one evening by the Prime Minister to an emergency meeting in the "Bunker", the headquarters of the War Cabinet situated deep beneath ground in the vicinity of Admiralty Arch. His journey by taxi which would normally have taken less than ten minutes, was considerably interrupted as it had to be undertaken at the height of a particularly fierce "blitz" by the Luftwaffe on the centre of London. Bombs were raining down, buildings were on fire, exploding anti-aircraft shells and flares dropped by the enemy lit up the sky. Many detours and forty minutes later, the cabbie leaned from his seat to open the rear door of the taxi to put Sir Philip down a few yards from his desired destination — "What a night!" commented Sir Philip as he paid the fare, topped up with a generous tip to reward the determination and courage displayed to get him there. "You're right, Guv, it is, but it could be worse — it could be raining" was the cabbies' reply. I recall Sir Philip saying with a spirit like that in our big cities there could be nothing but victory for our cause.

Apart from the concentrated air raids on our cities a large number of bombs and other explosive devices such as land mines were dropped in rural areas. In some cases it was difficult to imagine what they were aimed at and it is possible a few were dropped at random to spread fear over as wide a section of the population as possible with a view to breaking general morale. In some instances these attacks brought death and destruction, such as was the case in the small hamlet of Hatford where cottages suffered direct hits resulting in the tragic loss of lives as folk lay asleep in their beds.

Ann and I had the frightening experience of being involved in one such incident while visiting Anns' elderly mother at 180 Wootton Road, Abingdon, towards the end of March 1941. It was about 10.30.pm. when our "goodnight" exchanges in the hall with "Granny" and her single son Tim, who had just returned from an evening's courting, were interrupted by an eerie loud whining sound followed almost simultaneously by a tremendous explosion which seemed to lift the whole house six inches out of the ground. This process was repeated rapidly every other second or so during which a number of bombs were dropped by two raiders and meant to fall on Abingdon R.A.F. Station. Both raiders had managed to mingle stealthily and undetected with a number of our Whitley bombers who were engaged in "circuits and bumps" as part of flying training exercises. As a result no local air raid warning had been sounded, so there was no opposition to their attack whatsoever. Despite this, only the first bomb dropped in the target area and the rest fell on the Wootton Road housing estate, one house of which received a direct hit, while eleven others were extensively damaged. There were in fact some remarkable escapes as there were seven people in the house which received a direct hit, but only one, a nineteen year old youth, who was billeted there, was injured. He was having a bath at the time and the poor lad received a leg injury which necessitated the amputation of his foot. The bomb had fallen through the landing into the bedroom of the seven year old son of the tenant Mr Laing; miraculously, although he was in bed he did not receive a scratch. All this happened just next door but one and I freely admit I have never been so frightened in all my life.

It all happened so quickly that it felt unreal, I remember shouting to Ann to get into the cupboard under the stairs and trying to push Granny, who seemed to have stiffened with fright, under the gate-legged table in the middle of the sitting room, but having managed to get her head under the table, the more I pushed her rear end the faster the table moved across the room. It was at this moment when the explosions seemed to be getting nearer and increasing in intensity that the glass and window curtains flew across the room in ribbons and I really thought this is it. For a split second or so a strange feeling of acceptance came over me, which converted almost simultaneously to one of relief as I realised the repercussions from the next explosion had diminished slightly and that it was on the other side of the house. In fact the next bomb dropped in some allotments about

100 yards away on the eastern side of Wootton Road, bringing down a large number of overhead telephone wires as it did so. Venturing outside into the darkness which was charged with the pungent smell of cordite, I saw the light of a shaded torch belonging to the Air Raid Warden living nearby who was anxious to know if there were any casualties to report at number 180. I was able to tell him apart from shock and the windows having been blown in we were all alive and kicking. Almost immediately I was attracted by a tirade of well chosen but unrepeatable threats as to what someone would do when it came to his turn to get his own back on the Luftwaffe. It came from a few yards further down the road and was uttered by a young R.A.F. Officer who had inadvertently driven his Riley Alpine Sports into the mass of coiled telephone wires lying across the road. I found him half-way underneath the car tugging at wires which had become wrapped round the transmission shaft. He was obviously not in the mood for any sort of conversation so I returned to No: 180 which was obviously temporarily uninhabitable for someone of Granny's age, so she packed her bags for the purpose of staying with us at Baulking while the house was repaired. For security reasons brother-in-law Tim did in fact manage to stay there while this was being done. As we prepared to leave we were abruptly reminded the incident was not completely over and indeed but for the space of a few minutes just might have been worse. Our old Austin 16 was parked in the service road outside 180 and as we opened the doors there was a considerable smell of something smouldering which was explained by the fact that a nasty jagged piece of bomb splinter had penetrated the laminated windscreen slightly to the left of the drivers line of vision, struck the beading above the rear window and come to rest on the rear seat in which it had burnt a large hole. As it was impossible to obtain a new windscreen we filled the hole with putty and drove the car in this state of make do and mend to the end of the war, when I sold it to Claude Ryman, the

cattle dealer from Didcot for £150, the final amount of the deal being decided by the toss of a coin in the "The Boot" at Aston Tirrold.

The war years have been recorded from many different angles, by many well known and interesting people and is better left to those best equipped to research and write it all down. Suffice to add — seen through the eyes of a dairy farmer, on 170 acres of heavy loam overlying clay it wasn't easy to switch from all grass to growing fodder crops such as kale, cow cabbage, mangels and swedes, as well as sugar beet as a small contribution to the nations need for sugar and the provision of sugar beet tops for the milking herd. On our heavy land towards the end of October the crop was difficult to lift and put on rail at Uffington Station for the refinery at Peterborough, and the amount of tare in the form of soil was often quite high. It is certain the growing of these high labour demand crops would never have been possible without the help of the "Land Army girls and gangs of P.O.W.'s organised by the War Ag" and administered in the Abingdon, Wantage and Faringdon area by Sam King, a great character and farmer's friend, whose life at times must have been hell, when trying to keep everyone happy by producing Land girls and P.O.W.'s out of a hat at a minutes notice, not only for root growing and harvesting but corn harvesting and threshing. Combine harvesters were almost unheard of then. I think Dick Warburton of Shillingford on Thames, owned one he had more or less made himself in the mid 1930's. Dick was a man before his time in relation to mechanised farming. I remember it being said of him when he was being introduced by the Question Master of an Agricultural Brains Trust one evening, that he was the man who introduced the wheelbarrow to S. Rhodesia. I had the privilege of serving on the Berkshire Executive Committee of the N.F.U. under his admirable Chairmanship and was on the local committee of stewards for the World International Ploughing Match when it was held on his farm at Shillingford in 1956. The event was commemorated by the building of a cairn made up of stones bearing the name of every competing country, and was erected adjacent to the field in which the competition took place. I am almost sure the champion ploughman was home bred and came from Devonshire. I recall the service of dedication of the cairn by the Bishop of Oxford prior to the commencement of the match, which lasted three days, as being a very moving occasion. Sadly, I heard recently that the cairn had been vandalised and some of the stones brought by competitors from many parts of the world had been taken away. What satisfaction someone gets from desecrating a structure commemorating a rare and interesting event leaves me wondering what sort of people they really are and for what reason they wish to destroy even small parts of what go toward making up the richness of our heritage.

Looking back once more to the few weeks immediately following the declaration of war, it is an understatement to say the whole farming scene was literally turned upside down to bring it into line with the policies of the Ministries of Agriculture & Food, which were basically to make sure the land and the reduced number of those working on it produced the crops needed most for the nations larder during what was likely to be a long and desperate struggle. To this end a committee of well known and respected farmers and landowners was quickly appointed for each county and they in turn appointed district committees within their counties whose job it was to visit farmers, explain the Governments policy and assess how best within their physical and managerial capacity they could co-operate by ploughing up some of their grassland to contribute towards the nation's urgent requirements.

These committees were named as County & District War Agricultural Executive Committees and possessed statutory powers enabling the enforcement, if necessary, of any recommendations agreed with the farmer following an on-the-farm assessment. As a state of emergency existed the Ministry of Agriculture also had the power to dispossess in cases of outright non co-operation or complete failure to reach a reasonable standard of production. Fortunately, such action was rarely necessary.

The Government were in fact demanding a much larger arable acreage in order to maximise the wheat crop to provide our daily bread and in the case of dairy and livestock farmers to encourage as much self-sufficiency as possible, not only to provide more forage crops but to use the plough on worn out permanent pastures for the purpose of growing more and better grass.

I remember my mentor Freddie Cox who with all the staff of the Berkshire Agricultural Instruction Dept. had been hastily seconded to what became known to farmers as the "War-Ag" committees, asking if we would co-operate in a 3/4 year project on 10 acres of our lowest and wettest land for the purpose of demonstrating hopefully what could be achieved by re-seeding this kind of land which was typical of quite a number of acres in the Vale of the Uffington White Horse. I am ashamed to say, at that time the sward consisted of about 70% of what we called "bull poles" — these were large tussocks of rush grass measuring 10/12 inches in diameter. These did their best to crowd out the 30% of struggling permanent pasture and the whole ten acres provided summer grazing for about five cows, if you could get them to eat it.

Freddie managed to get Dr William Davies the well known grassland expert and deputy to the great Sir George Stapleton at the Welsh Plant Breeding Centre at Aberystwyth to come and advise us what to do.

He arrived at Church Farm one wet miserable day in mid January. The field was waterlogged and having struggled over it in his Wellingtons he said it was obvious in the long term it needed to be drained, but it was war time so for the time being that was out of the question, nevertheless he was certain a great deal could be done to bring it into a reasonable and acceptable level of improvement. His advice was to disc-harrow the surface as many times as possible, the wetter the better in order to allow the discs to cut up the tussocks, and then plough deep enough to bury the trash when the land dried out in the spring. The grass and clover mixtures would be planted preferably by drilling and not broadcasting by the middle of May; having attended to recommendations following a soil analysis, one of which was the application of 10cwt of burnt lime per acre, wicked stuff to spread. It came in 1/2cwt bags which soon burst open when stored due to absorption of moisture from the atmosphere. Being a very fine powder it got into one's eyes and up one's nose when loading and blew all over you and the poor old horse when spreading from a horse drawn fertiliser spreader. The only precautions available seemed to be choosing a quiet, windless day and tacking four or five sacks on a piece of wood and fixing to the rear of the spreader so that it trailed like a blind. Clearing the soil by an inch or two, it did help most of the lime to fall to the ground before it could become airborne.

The seed mixtures were designed to produce a ley of 3/4 year duration and were based on rye grass, timothy and white clover. Of the ten acres, a plot of three acres was sown with mixed strains of S.23 and S.24 rye grass and S.48 and S.51 timothy together with 1. 1/2lb S.100 white clover and 1/2lb Kentish wild white clover per acre all these except the Kentish wild white were strains bred and developed at the W.P.B. Centre at Aberystwyth. A similar plot was seeded to the same mixture but composed of seeds from commercial origin, while the rest of the field was seeded to a general purpose Cockle Park type mixture including a wider selection of grasses and clovers for the purpose of observing how they behaved and persisted on our heavy land.

Following a fair amount of struggling, getting stuck and unstuck many times while progressing from rushes to a reasonable seed bed — so much so, that Maurice Reade, a friend and neighbour, leaning on the gate to the field one afternoon while my steel wheeled tractor was slowly trying to dig itself out of sight, suggested that I needed my head seen to.

Nevertheless a reasonable spell of weather enabled us to get the seeds drilled towards the end of May and a successful "take" gave us some valuable grass for the cows in the middle of July at a time when the old permanent pastures had run out of puff and the aftermath grazing following haymaking had not yet arrived.

Following two or three further periods of "on & off" grazing which had beneficial effects both on the sward as well as milk output Freddie decided there was enough evidence available to substantiate holding a local demonstration on site one evening the following spring, using it as an occasion to put before farmers and their employees not only the urgency of increased production but useful ways supported by figures and graphs supplied by Arthur Foot from the National Institute in Dairying at Shinfield, Nr Reading. I remember it being a somewhat blustery evening and he had difficulty in securing the graphs to make-shift stands attached to board and trestle tables displaying weighed amounts of various cattle feeds etc.

I recall Freddie and his fellow organisers being satisfied with the turnout and interest shown by the fifty or so people present. The field was adjacent to the Stanford road so as the months went by it was easy to observe the project and draw one's own conclusions. Visitors were always welcome as they were also to a small number of other seasonal trials for the purpose of comparison and suitability of different varieties of wheat and oats. Barley was not then considered to be a particularly suitable crop for the heavier land in the Vale, but it is interesting to note some fifty years later due to the efforts of plant breeders and the improved conditions brought about by land drainage schemes as well as the arrival of the combine harvester that excellent yields from this crop on Vale land are now readily obtainable.

Like many dairy farmers faced with a much larger arable acreage, in our case we had not grown any arable crops for five years, the lack of cultivating equipment posed considerable problems. Some farmers faced with a ploughing-up order and without a tractor or plough had no option but to join a queue for the services of a contracting pool hastily set up and equipped by the War-Ag to alleviate such problems. Those with a tractor but little else, soon became aware of the possibility of borrowing equipment from neighbours and others when they were not using it on the basis of a small hire charge per acre to cover wear and tear. Almost everything one needed required a permit from a crawler tractor to barbed wire and the staples required to fix it, to the home-made stakes — wood was rationed too. Obviously the pressure on the clerical staff issuing the permits etc. was intense and in some cases considerable time elapsed before permission to purchase was granted. I remember hearing of the farmer who needed half a dozen rolls of barbed wire; getting so frustrated at not receiving acknowledgement of his application; sending a stamped addressed post card to the County War-Ag asking them to tick the appropriate reply and send it back to him by return of post.

The reply options he listed for them were:—
  a. We have received your application
  b. We have not received your application
  c. You can have some barbed wire
  d. You can't have any barbed wire
  e. We haven't got any b.. barbed wire
  f. To hell with you and the barbed wire

The returned post card selected d and e bracketed together and suggested read concurrently they provided not only the answer but the reason as well.

The demand for farm machinery quickly resulted in a desperate shortage which was thankfully alleviated to some extent by the U.S.A. who were sympathetic to our cause but at that time had not joined in the struggle, devised a scheme known as 'Lease Lend' which was a kind of hire purchase arrangement enabling them to send us a wide range of equipment and supplies in response to Winston Churchill's famous plea "give us the tools and we'll finish the job".

We were fortunate to be granted permission to purchase an

M.M. sixteen coulter, double disc, corn and seed drill for £80. It was manufactured in Minneapolis and shipped over the Atlantic under the terms of the lease lend agreement. As well as planting corn very well, it was excellent when drilling grass and clover seed mixtures, its double disc coulters enabling the seed to be planted at a minimum uniform depth of about 3/4 of an inch, where it could benefit from soil moisture and not have to struggle too long to reach the light. The drill became very popular with neighbours and friends, the hire charge of a shilling per acre providing a little beer money as well, though it must be said beer was almost in shorter supply than the cash needed to pay for it, a state of affairs borne out by "Regular Customers Only" notices which appeared in the windows of quite a few locals.

Looking back on the five war years I never fail to marvel at the outcome. Considering we started way behind scratch and for sometime stood alone, it is amazing, despite the setbacks of the first two years how quickly people accepted the gravity of the situation, became organised to cope and responded to every call made upon them to increase their part in the war effort. Coupled with the eventual backing of the U.S.A. following the Japanese attack on Pearl Harbour and Hitler's blunder in stretching his forces beyond their limit when opening up his eastern front by going to war with Russia, the tremendous support from our colonies overseas, not forgetting the courage of those in the Resistance carrying on the fight in great danger behind enemy lines in many parts of occupied Europe, all this resulted in the complete surrender of Germany & Japan on V.E. & V.J. days in 1945.

The relief was tremendous in knowing that the killing of soldiers and innocent civilians in many parts of the world had at last ceased, and that people could sleep soundly in their beds knowing once again that there was no need to lie awake wondering, hoping and praying that the aircraft they could hear over head was one of ours and not one of "theirs".

Celebrations on the evening of V.E.Day consisted mainly

of dancing in the streets, squares and market places up and down the country. As our first child Pamela Mary was only 9 months old, it was difficult to ask anyone to baby-sit on such an occasion as this so Mum stayed at home while I journeyed to Faringdon to celebrate on behalf of us both! The market place was thronged with people of all ages singing and dancing around a huge bonfire and people were spontaneously being lifted aloft and chaired once round the bonfire to the echo of cheers from everyone present whether they knew them or not. It was all unrehearsed of course and so simple but the enjoyment shown on peoples faces said it all. It went on until the small hours and I returned home to tell Ann what fun it was and that I had never before danced with so many different people during one evening — which perhaps was not the most sensible thing to say but under the circumstances she understood.

As it was only twenty seven years since the end of the First World War, memories soon began prompting people to say "we have won the war, let us make sure this time we can win the peace and avoid the misery and depression of the 1920's which followed the last one".

There was of course a mammoth task ahead to rebuild cities and to get industry and food production going at a pace to satisfy the demand there was bound to be for manufactured goods of all kind as well as to bring about the lifting of food and fuel rationing as soon as possible.

It was only natural that individual industries were beginning to jostle for priority in the overall scheme for things to come. So many post-war committees burnt the midnight oil discussing how best to approach and deal with the many problems facing their own particular interests for the future. In this new climate The National Farmers Union was not alone in looking for ways and means to do what it could for its members, which probably constituted at least 90% of the farmers in Great Britain, operating under its motto "Unity is Strength". My family had been members of the Faringdon Branch since its foundation in about 1921, Grampy George Ernest being an early Chairman, as were Uncle Pete, and my father in the mid thirties. In 1949 I was elected to follow Ernest Pearce of Kilmester Farm, Eaton Hastings and was in turn succeeded by a very great friend Victor Tytherleigh who in the mid thirties came from Shaw in Wiltshire to farm at Northfield Farm, Faringdon.

It was not long after this, while reading a book which I think was entitled "Rich in My Heritage" and written by a country-lady whose name I cannot recall that I came across these three verses of poetry which I thought were worth copying and keeping in mind.

They were written at intervals during and after the First World War:—

1914

> Worthy farmers of Britain, hardy yoeman all,
> Save this land from hunger, hear your country's call.
> Now that the war's dark shadow threatens our daily bread
> Plough up your choicest meadow, sow it with corn
>   instead.

1918

> Danger is still approaching everything is at stake,
> Put forth your utmost effort, fear not the risk to take.
> England has learned her lesson in tears and grief at last,
> Ne'er shall a grateful country slight you as in the past.

1921

> Once more the seas are open, once more the skies are
>   clear
> Gone is the fear of famine. Why is the corn so dear?
> You farmers made a fortune out of this beastly war,
> Now you can go to the devil, just as you did before.

(with acknowledgements to the author W.H.Still)

I have recited these lines quite a few times on appropriate occasions one of which was quite recently on the B.B.C. Radio programme "Farming Today" — in concluding my recitation I acknowledged the composer as being a Mr W.H.Still, adding the remark "whoever he was?" which prompted a listener named Mr D.A.Young of Blackgrove Farm, Lingfield in Surrey to write and tell me he knew him very well and that he was a very popular local character who lived at Addington Court Farm, near Croydon, before he died some twenty years ago. He also mentioned that he was a very good after-dinner speaker and was present when he made his very last speech at the age of 93. In fact I received other letters following that broadcast requesting copies of his verses; so I guess his name will be remembered by others for a few years to come. I am grateful that Mr Young took the trouble to inform me of his identity, thus adding considerable interest and meaning to the verses I have kept by me for the last 50 years.

NINE
Post War Period

I think the post war years up to around 1960 became a period of purposeful endeavour by almost everyone. Motivated no doubt by feelings of relief and the spirit of comradeship which had developed during the desperate national struggle to survive. Certainly in farming everybody seemed to want to help each other. Maybe we were heeding the words of that wise man of Kelmscot Manor, William Morris 1834 — 1896, who wrote "Meanwhile, let us not sit around in darkness, like fine gentlemen; but rather by some dim candlelight seek to put our workshop in order in readiness for tomorrow's daylight".

With this mood prevailing, it was rather like a period of reformation during which a large number of Farming and Growmore Clubs sprang up all over the country, conferences were organised in abundance, mainly by the Min. of Ag, on all aspects of farming and its future. Many Commercial Companies such as I.C.I. and B.O.C.M. were quick to follow, opening up their factories, mills and experimental farms and providing film shows etc. on all sorts of occasions. N.F.U. meetings at local, county and national level were very well attended. At that time almost every market town boasted a branch of the N.F.U. where grass root members could air their views and following debate despatch any worthwhile resolutions to County Branch level in the earnest hope that further debate would gain sufficient support to enable the matter to proceed to headquarters in London where the President and Council members were always in close touch with leaders of Government departments dealing with all matters concerning farming and food production.

There was also understandably, a close liaison between the N.F.U and the M.M.B. brought about by the membership of Regional committees of the Board being comprised mainly of nominees put forward for selection by N.F.U. county milk committees. I did in fact have the honour to serve as one of the three Berkshire representatives on the Southern Regional Committee of the M.M.B. from 1955 to 1972 and am a tiny bit proud to know that my son Bill, now farming close to Basingstoke, represented Hampshire on the same committee when the M.M.B. was disbanded on November 1st, 1994.

The Southern Region of the Board embraced Buckinghamshire, Oxfordshire, Berkshire, Hampshire and the Isle of Wight and during my stint of service I was fortunate to meet regularly many interesting knowledgeable farmers from those counties, as well as attending combined Regional Conferences at least twice a year in London and listening to others putting particular points of view which didn't always fall in with our own way of thinking.

I was serving quite a while during Sir Richard Trehane's Chairmanship of the Board and came to marvel at his deep knowledge of the organisation and the positive leadership he displayed for the benefit of all milk producers, large and small alike.

I cannot proceed further without reflecting upon the mixed feeling of joy and sorrow I experienced in the short space of three years from 1944 to 1946.

In June 1944 Ann and I were blessed with the birth of our first-born, Pamela Mary, which I recall celebrating with one or two others by de-corking a bottle of very old port bearing an almost indecipherable label indicating it was of 1854 vintage. It came into my possession during a Red Cross Sale, to which it was generously gifted by Bunny & Queenie Taylor, mine hosts of The Crown Hotel, Faringdon. I think I gave a fiver for it. Due to its crust it was difficult and tedious to pour but not to drink and I can almost taste it now.

Pam was delivered at Church Farm by Dr. Pulling, ably assisted by Nurse Speed, our district nurse who lived at Stanford in the Vale. She was a great character and workaholic spinster who trundled around the Faringdon area day and night it seemed in her Morris 1000, becoming known to all of course as "Speedy". She could be a bit terse with the odd patient who perhaps stepped out of line by not adhering to her advice and instructions, nevertheless she was a dedicated carer and I'm sure ours is not the only family in the district to remember her with gratitude.

Pam was born in our bedroom at Church Farm in the early hours of a mid-June morning by the light of two "Aladdin" paraffin lamps, electricity not having reached the village by then. The "Aladdin" lamp gave a light similar to gaslight and was one of very few that incorporated a mantle covering quite a small round wick needing very fine trimming, which when achieved produced a light far in advance of the light given out from the bare flame, coming direct from a wick impregnated from paraffin contained in the lamp bowl below. The Aladdin, it must be said, was prone to misbehaving when, on leaving the room for an hour or so, the wick was turned down low and the lamp happened to be standing in a slight draft. The mantle would then soot up into what can best be described as a glowing cinder about the size of a shuttlecock, turned upside down; this procedure was accompanied by a strong smell of burning and I imagine a distinct possibility that the whole thing could burst into flames. I remember George Heavens who once lived at Cheivley House, Little Coxwell, saying that he arrived home late one evening, his wife having gone to bed, had turned the Aladdin down and as he opened the sitting room door was almost choked by the smell and scared stiff by the dull red glow coming from the lamp. Despite the fact that the ink was hardly dry on the cheque he had just paid to have the room decorated he grabbed the stem of the lamp opened the window and hurled the whole thing, glass, shade and all as far across the lawn as he could get it. Without further ado, having seen himself upstairs by candlelight, feeling very cross and no doubt bent on apportioning some of the blame for what happened upon his wife. He awoke next morning to suffer his anger rekindled by the sight from his bedroom of the smashed "Aladdin" in the middle of the lawn, only to be astonished and relieved a few minutes later as he opened the sitting room door to find that apart from a lingering smell of burning soot there was virtually no visible damage at all. One can only conjecture what might have happened.

Ann's confinement was lengthy and difficult. As well as having lasting memories of her courage, I must confess it was the first time I had experienced *real* anxiety. Even now it is beyond me to describe the depths of my feelings at the time. The number of times I walked the hundred yards from Church Farm to the railway bridge and back that summer night I'll never know and when Dr. who had been on site with nurse since about 10.30p.m. said to me at 6.a.m. — 'I'm getting my partner Dr. Graham Stenhouse to come out from Faringdon, I need some help' — my feelings touched rock-bottom. However abundant joy followed not long after Dr Stenhouse's arrival, when at 8.30.a.m. precisely Speedy shouted downstairs to me — "Its all over — a baby girl — see you in a minute or two". When she did appear I gave her one of the biggest kisses and hugs I have ever given anyone and I also remember my words of gratitude towards the two Doctors expressed in the farmhouse kitchen over a cup of tea, seemed totally inadequate.

Sadly this period of joy was to receive a severe jolt as towards Christmas my mother's health took a turn for the worse which resulted in her being admitted to the Radcliffe Infirmary, Oxford, where she died from a coronary thrombosis on 2 February 1945 at the age of 63. It was indeed a great consolation to know that she had at least seen and held her first grandchild in her arms, a sight I shall never forget. There was no doubt we were all going to miss her a great deal, especially father. To me she was a kind, loving mother blessed with a great deal of common sense, abhorred any form of dishonesty, was a regular church goer and had many friends.

It was only a few months after mother's death that father told me he intended conveying Church Farm into my ownership by a Deed of Gift explaining that this fulfilled a desire both he and mother had shared for sometime and he felt this was now the appropriate time to do it.

The gratitude Ann and I felt for their joint loyalty and the spur it gave us for our future deepened a feeling that mother might have had a premonition of some sort of difficulty and was anxious to do what she could while she was alive.

We shall never know; but after a while father married again. Alas, the marriage was a complete failure, despite our earnest efforts to reconcile two very different characters. It was my first close experience of a breakdown in a serious relationship and I couldn't understand why two people who must have had deep feelings for each other over a not inconsiderable period of time were able to sit round a table and seemingly with help, resolve their differences one day and fight like tigers the next. It was a sad interlude during the early part of my own married life but I like to think it probably taught me something about a side of human nature with which I had not previously come into contact. Suffice to say the problem was settled by an agreed separation.

Once again a period of sadness was considerably relieved by one of joy surrounding the birth of a son and heir, William John, on 4 July 1947. To have been blessed with a "pigeon pair" was a great thrill. We would like to think the fact that we referred to him as "Bill" long before he was born had some effect — but I doubt it. The reason for "Bill" was that I was known by that name at school — think it was something to do with being a farmer's son. A kind of 'Ol Bill the farmer image, arrh — arrh. I am pleased to say Ann had an easier confinement this time ably carried out at Church Farm by the same team of Dr.Pulling and Nurse Speed.

Not long after our happy event I was motivated by listening to Mr Chambers, our local auctioneer and secretary to the

Faringdon Branch of the N.F.U. He was making the point at one of our meetings that farmers should be prepared to get off their backsides and put themselves forward for election as parish representatives on Local District Councils if and when the occasion of an election arose. He supported his plea by saying that farming and its ancillary trades represented the main interest in most villages and it was for the good of the community that their voice should be heard.

It so happened that elections for the Faringdon Rural District Council were due to take place in the spring of 1947, so I made up my mind to do something about it as the previous representative did not wish his name to go forward for re-election.

Baulking being a very small parish with barely seventy people on the Electoral Roll, its single representative on the R.D.C. was usually returned unopposed. However, such matters were destined to take a different course following a visit one evening by a neighbour Tom Saunders, the local School Attendance Officer. Over a bottle of beer in the sitting room at Church Farm he explained that he was willing to represent the parish, he needed a proposer and seconder and would I propose him? "Well Tom" says I, " that's impossible as I was hoping you would do the same for me, looks like we are in for the excitement of an election in the parish". Tom accepted the position, so we had another bottle of beer, laughed, shook hands and wished each other the best of luck. Consequently, the "battle" took place on the 31st day of March 1947. I had a gut feeling I might just scrape in but would not like to bet on it. In fact in an almost 100% turn-out I polled three more votes than Tom, close enough he felt to call for a re-count, which was duly carried out in front of us both, with the same result and most importantly no loss of friendship between us.

Having been elected to participate in matters and schemes at district level for the betterment of the parish I felt priority number one was to obtain a supply of electricity for the village.

Early correspondence with the Southern Electricity district office at 33 West St Helen's Street, Abingdon revealed that "due to the large amount of priority work on hand and the difficulties of obtaining supplies of materials, it was anticipated that it would be 12 to 15 months before the work could be completed". So it was not until March 1949 that the S.E.B. completed the scheme for the extension of mains for providing a supply to the village which would include the farms, farm workers cottages and other premises situated adjacent to the Village Green. This was confirmed in a letter dated 26 March from a Mr Rushby the District Executive Officer of the S.E.B. confirming that he was happy for me to place his proposal before our Parish Meeting. Not long after that we were able to put away the candles and oil lamps and replace some of the petrol and diesel engines with electric motors for the purpose of operating milking machines, heating water, providing outside lighting, domestic uses and many other of the benefits of electricity which we accept today as a matter of course.

It wasn't long after this, that someone at Whitehall had the idea that local government needed to be re-organised for the purpose of stream-lining the existing system to make it more efficient and so it was said — to save money.

The outcome as far as our parish of Baulking was concerned was that we were amalgamated with Kingston Lisle, a parish with considerably more than double the number of electors than us and already represented satisfactorily by Captain Prioleux. Despite our protests at the loss of being represented by one of our own parishioners at Rural District level and the almost certain failure of contesting the seat, there was no other option, but to lobby the Captain intensively when necessary. When he decided to retire he offered to support me in the event of an election should I be prepared to stand. In fact the situation did arise and I have no doubt my election to serve both parishes owed a great deal to the support he gave me. I continued to serve in that capacity until in 1977 someone else in Whitehall suggested further re-organisation along the same lines as before and for the same reasons with the result that Faringdon, Wantage and Abingdon Rural District Councils were made redundant — our part of Berkshire became part of Oxfordshire and we were now to be governed locally by The Vale of the White Horse District Council, with offices situated in Abingdon. The council now consists of an elected member to represent about six parishes in some instances. I often wonder if anyone has ever worked out if the upheaval really has saved money and worked as efficiently as we were told it would, also if the villages, especially the smaller ones get the same representation that was provided by someone who actually lived in the village and was therefore much better aquainted with its conditions and possible problems.

Up to now I have done my best to describe roughly the first half of my life. Being now in my eighty fifth year I have to remind myself that if I don't gallop on a bit I might not be able to complete — I think it was Sir Winston Churchill who once said "There is no man in a hurry like an old man"!

In 1951 the accumulation of figures resulting from fourteen years participation in the costing of milk production and farming profitability conducted by Reading University on a varied sample of farms in the Southern Region of the M.M.B. covering Berkshire, Oxfordshire, Buckinghamshire, Hampshire and the Isle of Wight I found myself in the vulnerable position of being invited to take part in a Conference being organised by the Berkshire Agricultural Executive Committee, the subject for discussion being "Milk Output per Man". It so happened at that time we were emerging from a few beneficial effects of a mini-revolution at Church Farm. Following a great deal of deliberation and my father's helpful attitude in letting me have sufficient rein to bring about change in our farming methods and most importantly to support the financial approach to our bank who, incidentally, to my certain knowledge have "accommodated" (if that is the word!) at least five generations of Liddiards.

The Conference, an all-day affair, held at the Corn Exchange in Newbury was attended by around 200 fellow farmers. It was chaired by a Mr J Edwards who was I think Chief Executive of the M.M.B. Among others taking part were Tom Bucknell, a chap named Pasfield who was the Min. of Agriculture's building expert and Richard Pettit-Mills who summed up.

Being the first time I had held forth in public, I found it a very daunting experience.

The following autumn I was invited to take part in a Milk Production Conference organised by the British Oil & Cake Mills at Harper Adams Agricultural College at Newport, Salop.

The main speaker was the current Chairman of the Milk Marketing Board Mr Tom Peacock, CBE, who spoke on the future of milk production in the U.K. and the Boards' marketing policy. B.O.C.M. being one of the largest compounders of animal feeding stuffs and with an obvious eye to business followed this up by inviting two of its many customers in the persons of Mr Boddy of Holme Farm, Caunten near Newark (100 acres) and myself (167 acres) to describe to the 450 dairy farmers present our efforts to provide enough money from cash crops to buy in feeding stuffs to secure the maximum profit and the greatest food production from a reasonably small acreage. We were both very much of the opinion that to strive for too much self-sufficiency our food production would be less and our farming less economic. We felt this view was borne out by the relevant costing provided for us by Nottingham and Reading Universities, with whom we were co-operating at home.

Ann and I were among the guests entertained to dinner the previous evening by the college Principal Mr Bill Price and his wife and it so happened that H.R.H. Princess Elizabeth, then heir to the throne had recently visited the college. Mrs Price was showing us a collection of photographs taken during the visit and I recall her pride in showing us her 'favourite' which recorded the Princess strolling through the rose garden with her husband, looking inquiringly at him following his description of the system of management concerning the out-door pig rearing unit they had just seen, which motivated Her Royal Highness to pose the question "What level of mortality rate per litter are you recording Price"? The expression captured on Bill's face portrayed considerable surprise to say the least and he explained to us that for a brief moment he stumbled to find the answer; which I think he said was in the region of 10% of all piglets born.

It is I think somewhat strangely coincidental that some forty-four years later my two grandsons became students at Harper Adams.

A year later I received an invitation to take part in the Annual Oxford Farming Conference. The invitation came from the organising secretary Mike Soper, head of Oxford University's Dept of Agriculture. The Conference Committee consisted of Roger North (Norfolk) Chairman, W.H. Cashmore (NIAE), R.N. Dixey (Agric. Economist, Oxford), M.Griffith (Wales), Clyde Higgs (Warwick), J. Mackie (Scotland), R.M.Older (Kent), John Rowsell (Hants), five of whom being well known farmers and the others experts in various fields closely connected with the agricultural industry. It was the Seventh such conference and was staged at The Playhouse Theatre from January 5th to the 7th in 1953. Following an Inaugural Dinner at The Randolf Hotel attended by a number of dignitaries plus 232 members. The next two days were spent listening and discussing such papers as "Management Problems on the Mixed Farm", "Labour & Machinery Management" and "Livestock Production on the Mixed Farm". Papers were read by farmers from Hampshire, Aberdeenshire, Yorkshire, Glos, Herefordshire, Berks, Bucks, Pembrokshire, Cambridgeshire and Somerset. Buffet suppers rounded off each days proceedings, while numerous decanters of port, all by kind courtesy of Worcester College, enabled further discussion on a much more informal level to take place well into the night. I can remember Dick Roadnight being questioned about ventilation of his portable out-door dome-shaped pig huts during hot weather. He answered by saying it was automatically controlled by the weather as the semi-circular rear panel was made of nine inch elm boards which shrunk in the summer when dry leaving gaps between the boards and swelled during the winter thus closing the gaps and preventing through draft at that time of year! I also recall someone moaning about the design and shoddy workmanship of some of the combine harvesters that were currently coming off the production lines, suggesting that a fully trained monkey should accompany each one for the purpose of being able to get to the parts human beings were unable to reach (that was before the Heineken advert came into being!). He also thought it would be a good idea to trail a large magnet behind machines to pick up the bits and pieces which fell off them due to poor welding and nuts and bolts not being properly tightened before leaving the works.

It is now well over 40 years since I stood on the stage of the Oxford Playhouse with my knees knocking to the tune of a woodpecker pecking a hole in a tree while being introduced by Prof. H.G.Sanders, a truly likeable man who was then Professor of Agriculture at Reading University and had recently become Chief Agricultural Advisor to the British Government. I feel it might be worth appending my modest contribution as printed in the official report of the proceedings in the hope that it might illustrate a little of what we were trying to achieve at Church Farm as a result of the change in our method of farming in order to expand, to bring a little more grist to the mill; while at the same time share our experiences with others for the purpose of listening and learning from any comments they might make.

Wednesday January 7th, 1953
THIRD SESSION
Chairman: Prof. H.G.Sanders, Ph.D.
LIVESTOCK PRODUCTION ON THE MIXED FARM
by R.E.J. Liddiard, Esq.
Church Farm, Baulking, Faringdon, Berks.
Firstly, may I say that I am only an ordinary dairy farmer and that I depend entirely upon farming for a living for my

family and myself. Being an owner-occupier any capital needed for new plant or buildings has to come out of profits.

I have been co-operating since 1935 with Reading University Agricultural Economics Department in their annual milk farm costing surveys. The resulting accumulation of facts and figures have many advantages and one big disadvantage. The disadvantage being that all the recorded data, can, I find, be used as evidence to get one into an uncomfortable position on a public platform.

As I'm not very accustomed to speaking in public I'm afraid the punishment is now yours as well as mine.

Now, Mr Chairman, I think it is obvious that the economic production of milk is really the outcome of intensive and detailed planning all round.

This planning seems to me to fall into *two parts* — first: cows and crops, to get the greatest possible output, and second: suitable building layout and organised labour to deal economically with that output.

I have a feeling that the consuming public will refuse to pay much more for their milk, and if prices rise, sales will drop. This possibility coupled with the high price of feeding-stuffs, high wages, in fact high everything, could soon led us into queer street. It follows then that any successful expansion in production must depend upon our ability as dairy farmers to reduce production costs, so that we can sell milk to the consuming public at a reasonable price and keep their valuable custom and goodwill, perhaps even, to compete in the manufacturing market, for the production of more butter and cheese — or who knows? perhaps to offset what might happen should our precarious National economy come unstuck.

It is on these lines, Mr Chairman, that I hope to try and describe how we are striving to make the best of things down our way. I say we because in talking of one's farming activities the word 'we' includes the staff, on whose willing co-operation so much depends.

We have a modest plot of 167 acres in the Berkshire Vale of

the White Horse, on which we keep commercial Ayrshires — rainfall 26 in. The soil is a clay loam overlying gault clay, giving us a late spring and early winter, and up to 1940 the farm was all grass, of the old permanent variety.

When the first ploughing-up order arrived, we felt that ley farming in some form or other was the only way by which we could hope to retain, and perhaps increase our dairy stock numbers.

It was obvious for our type and size of farm, that milk production must continue to occupy the centre of our new farming system. During the war years of course, urgent necessity and national requirements dictated our plan of campaign, as they did for everybody.

In 1945 having taken advantage of the Government grant towards a water supply for every enclosure, we were able to draw up a comprehensive rotational cropping and utilisation plan for each field. Although this plan gets knocked about occasionally — not least by the weather — we try and stick to the fundamentals and so far we seem to have 'got by'.

I think it can be said that the six years that have passed since then have been a fair test from the weather point of view, and our production in terms of stock carried and milk produced has increased as you will see from Table II.

To follow a seven course rotation we split the farm into eight blocks — seven rotational blocks of approximately 19 acres each and one block of 34 acres of permanent pasture consisting mainly of passage and access fields. As you will see from Table I, the original rotation was two wheat crops followed by spring oats and tic beans, followed by four years ley. Two years ago, however, we modified this rotation a little in order to avoid the difficulty we were experiencing on our land, of getting the spring corn planted early enough. To this end we shortened the arable break to two years and brought in winter oats on half the block for the first year. We then follow with wheat and a five year ley. So we now have each year:—

10 acres	Winter Oats
28 "	Wheat
95 "	Leys

Plus the produce of the 34 acres of permanent pasture, approximately one-third of which are cut for hay each year. Now Mr Chairman, before I describe how the produce of these 167 acres is turned into milk, may I explain our attitude towards self-sufficiency.

TABLE 1

Cropping Plan	7 blocks of 19 acres	133 acres
	1 block perm.past.	33
	Total	166 acres

Old Scheme	New Scheme
Wheat	½ Winter Wheat
	½ Winter Oats alternately
Wheat	Wheat
Oats & beans (spring)	5 year ley
4 year ley	

Oddly enough, as I mentioned at the start, we are trying to farm for a living, and having done a good deal of hard thinking, we feel certain that, to become too self-sufficient on a farm of our size would mean a considerable reduction in the number of cows we could keep; and although we have figured that we should be feeding them a little more cheaply, we are convinced that the gross output and profit from the farm would be considerably less. I make no apology for mentioning this, as I feel it is a matter of great importance to all small and medium sized dairy farmers.

With this fact very much in our minds, we decided that our policy would be to *produce as much of the cheapest form of maintenance fodder as possible for the winter,* so that we can keep the greatest possible number of milking cows, and rear all our own replacements and to strive during the summer to make a fixed grazing acreage produce maintenance and three gallons for that number of cows, as well as maintain the followers. For the production ration in the winter we purchase as much concentrates as we are allowed, and supplement them by dried grass cubes, either purchased or dried on contract from odd plots of our own grass when, and if, available during peak growth periods. Having said that, Mr Chairman, I return to crop utilisation: the wheat obviously provides useful cash for purchase of concentrates, and the straw is needed for litter. Some of the winter oats are retained for feeding to the young stock with a little purchased protein, the rest being sold. Of the 96 acres of leys, 38 are grazed, only 19 early grazed and late hayed, 38 cut for hay and aftermath grazed. This hay acreage plus part of the permanent pasture, with the generous use of fertilisers, usually gives us somewhere near 120 tons. Taking the winter feeding period at about 165 days, from mid-October to the end of March, we require about 74 tons of hay for the 50 cows, allowing 20 lb. per head per day, or in our language, one bale between 3 per day. That leaves around 40 tons plus the oat straw for the followers.

All the leys are sown direct, and each individual ley block managed and utilised in the same manner, rotationally, from the time it is sown until it is ploughed out. I have tried, Mr Chairman, to give a brief description of the methods we are using to produce the milk, and I would like now to say a word on our building and labour reorganisation, and the minor revolution that took place in 1948 to enable us to become accredited T.T., and finally attested.

It was obvious that the successful accomplishment of our task would not only earn us valuable premiums (£600) but at the same time our labour operating costs *must* be kept down to a minimum. As the rebuilding of existing sheds was impracticable, we had *two* alternatives.

A. Building a new cow-house incorporating a dairy and food store for fifty cows at a cost of just over £3000. *or*

B. Concreting existing yard and converting old carthorse stables into a milking parlour. The stables were no longer needed as such, as the last 'old faithful' had gone. The existing sheds with tie-ups for 52 would continue to house the cows in winter as before. The estimated cost of this conversion was £1900, the installation of an Auto-Recorder being included in that figure.

After careful consideration this last project seemed to be the better, for the following reasons. It was cheaper by over £1000; it would, we hoped, enable us to adopt a more economic system of production and allow us to reorganise our small labour staff with advantage — provide better working conditions; and lastly, most important of all, it would not set so severe a limit on the expansion of the size of the herd, as a partly covered yard adjoining the layout could accommodate stale milkers without extra cost.

The existing old sheds fitted in quite well with the proposed new layout. The old part-cobble-stoned cow yard an area of 1259 sq. yds. was concreted at a cost of just under 10s. per sq. yd., and divided approximately in half by a fence and gate, for assembly and dispersal. The dispersal yard was then divided in half again by another fence and gate; enabling both areas to be increased or decreased when all the cows are together at the commencement of milking during summer months, and decreased as milking progresses, to allow more room for dispersal. Also in winter when cows are housed, the sheds being on three sides of the dispersal yard, each shed can be let out, milked and tied up separately, with the cows grouped as nearly as possible according to yield, this enables the heavier milkers to be milked first in the morning and last at night.

With regard to the old carthorse stables which are now the parlour and dairy, fortunately there was an old hay and chaff loft running the full length of the building and this has made an excellent food store into which all feeding-stuffs are hoisted on arrival at the farm. The feed hoppers have been extended to the floor of the loft and each holds 1/2 ton of concentrates, approximately six days requirements at peak production, and is divided in half to save having to pre-mix the two types of concentrates being fed. This arrangement effects quite a saving in labour.

Another addition to the layout has been the conversion of two bays of a cartshed into two bull boxes, with exercising yards and service pens and bulls can now be managed with safety and the minimum of labour.

TABLE II
Comparison of results in 1945-46 and 1951-52

	1945-46	1951-52
Output per acre	£24 0 0	£56 6 0
Livestock units/100 acres	36.8	53.0
Output/livestock unit	£58 0 0	£85 5 0
Acres Feed/livestock unit	2.7	2.0
Output/acre of feed	£21 0 0	£40 2 0
Milk yield/cow	670 gals.	864 gals.

Note. Livestock units are arrived at on the basis of approximate feed requirements for each kind of livestock. The feed acreage per livestock unit includes the actual acreage of feed crops and grass plus an allowance for purchased foods at the rate of one acre for each ton purchased.

Now, Mr Chairman, a word on labour utilisation. The regular paid staff consists of three men, a cowman, dairyman, and tractor driver, plus a part-time man, employed mostly on maintenance and repairs. On alternate week-ends the tractor driver becomes cowman, and I take on the role of dairyman, so each of us has every other week-end free.

The actual milking and the cleaning and sterilising of the three point recorder is the sole responsibility of the dairyman, as is the feeding of concentrates according to a list posted weekly on each hopper. The cowman is responsible for all other work in connection with the cows such as fetching and taking, setting electric fences, etc., in summer; and in winter when cows are housed the feeding for maintenance, rearing of calves, cleaning out sheds while cows are being milked, at which time he is assisted by the tractor driver so that the dung can be loaded direct into a spreader to complete the job, so long as the state of the land permits. It is perhaps interesting to note that we spend 95 man hours per cow as against a group average, covering 45 farms, of 145 hours and 12 man hours per 100 gallons, as against the average of 20 hours.

As for the men themselves — we always have a small conference before any alteration in method or routine is made. Grumbles, if there are any, as well as suggestions are welcomed, as a contented worker is usually a good worker, especially if he feels part of the scheme of things, knowing what he has to do and how he is going to do it.

In conclusion, Mr Chairman, I give a few figures of comparison between a year of production of ordinary milk, with a bucket type plant and old layout, and a year with milking parlour and new layout producing T.T.milk. You will see them set our in Table III. For the purpose of defining a man, it has been taken that 1 man=48 hour week; for example, 1 man doing 60 hour week is, for this purpose 1. 1/4 men.

TABLE III
Labour Utilisation   1 man=48 hour week

Old System 1947		*New System 1949*
42 cows 2.4 men		41 cows 1.4 men
Output per man	11,366 gals.	21,090 gals (86 per cent inc.)
Dairy work	5,933 hours	3,508 hours
A saving of 2425 hours		
Labour cost	4.6d gal.	3.2d gal.
Per cow	£12 12s. 5d.	£9 14s. 0d.

The considerable saving in labour achieved has allowed us to expand our herd to its present size of fifty-two cows and sixty-seven followers, to keep in step with the gradual increase in the productivity of our soil and make more time available for the extra work created thereby, as obviously one can't have increased production without increased work — contrary as that may seem to the beliefs of man! Also there is little point in saving man hours if the time so saved is not turned into further good account.

Looking at the financial saving involved, it could be said that the saving in labour of £3 per cow which amounts to £120 on 40 cows, is more than paying the interest on the capital outlay. Charging 5 per cent on the £1537 spent on building alterations and 10 per cent on the £300 for the auto-recorder, the total charge for interest would be £106 17s. 0d. per annum. Deducting this amount from the amount of £120 we re saving, we shall still have a balance of £13 13s. 0d. on the right side. One must also consider the income tax relief obtainable on improvements of this nature.

As to the future, Mr. Chairman, I feel certain there will always be a demand for quality milk offered at a reasonable price to the consumer. This fact should give us the necessary confidence and encourage us to scheme for greater *output per cow, per man,* and most important of all, *per acre,* for the nation's sake as well as our own.

Following the close of the Conference on the last day, there was an interesting little tail-piece as far as I was concerned. It was rumoured that one or two members of the BBC Radio production team responsible for presenting "The Archers" — "Everyday story of country folk" had been in attendance. Late that afternoon I received a brief telegram from the producer Godfrey Baseley saying "Advise listen Archers to-night". Consequently, glued to the wireless, about half way through the programme while Dan & Doris Archer (Mum & Dad) were nattering in the farmhouse kitchen at home in Ambridge, the sound of a door opening and footsteps coming down a flag-stoned passage prompted Doris to say "Wonder who this is?" Whereupon Dan said, "Oh, I expect thats Phil (their son) returned home from the Oxford Farming Conference". The dialogue which followed Phil's arrival in the kitchen concerned a broad description of the proceedings and some of the impressions he gained. I think they are best illustrated by the following relevant part of the original "Archers" script, Episode 516 by Edward J. Mason and Geoffrey Webb, edited by Godfrey Baseley, produced by Tony Shryane and transmitted on the 9th January 1953:—

DAN:1. So you don't reckon you wasted your time, then Phil?
PHIL:2. Not on your life, Dad.
DAN:3. It all sounds very interesting from what you've told me so far, anyway............
PHIL:4. It was, but you know Dad, it's difficult to put your finger on the real benefits of a conference, but there are a few things that seemed to stand out in my mind, and well, I shall think of them again and again.
DAN:5. Such as what?
PHIL:6. Well, I think the debate was one thing.
DAN:7. What was that about?
PHIL:8. Whether the community is best served by smaller rather than larger farming units.
DAN:9. And what won that?
PHIL: 10. The smaller farm.
DAN: 11. Bet that shook you a bit, didn't it?
PHIL: 12. Frankly it did. And another thing I remember — the insistence of people on costing. Mr Batten from Gloucester asked how many farmers could really tell just what it cost to produce a cwt of potatoes — or a ton of barley; and then there was a Mr Liddiard from Berkshire who gave us the costing of a milking parlour as compared with the old type cow-house.
DAN: 13. Was there much difference?
PHIL: 14. Yes, very nearly double his out-put per man. Wait

	a minute, I've got the figure he quoted in my pocket. On the old system in 1947, for 42 cows he had 2.4 men, on the new system in 1949, for 41 cows — 1.14 men.
DAN: 15.	What d'you mean 1.14men?
PHIL: 16.	Well, he puts one man equals a 48 hr week.
DAN: 17.	Oh, I see. These costings are a bit beyond me.
PHIL: 18.	But you've got to know Dad, if you're going to make the job pay. For one instance he told us that on the old system his labour cost him 4. 6d. per gallon, whereas with the milking parlour only 3. 2d. per gallon. Well, if you work that out you save 1¼d. on every gallon produced.
DAN:2.	Did he say anything about man-hours working it all out in the office?
PHIL:3.	Now look, Dad, its no good you trying to dodge the issue. There's bound to be more office work of course, but I reckon it's great to see farmers coming voluntarily to a conference of this kind to try and increase their own efficiency. You better come with me next year.
DAN:4.	I doubt whether it would stick, alright for you chaps. Ah, there's Tom. Hey up, Tom lad. Top of the morning to you.........

While researching the correctness of the relevant script for "The Archers" programme with the BBC Radio Archives Department, I had the opportunity of speaking to Bob Arnold who at the age of 86 years still plays the part of Tom Forest the Ambridge game-keeper. Until fairly recently he lived in Burford but sadly after losing his wife he has now moved south to live with his daughter in Hampshire. He told me that he first broadcast for the BBC in 1937, contributing a collection of short stories concerning the Cotswolds, such as the public hanging of three highwaymen from a gibbet in the village of

Fulbrook, near Burford.

He was a member of the original Archers cast when it first went "on the air" in 1951 and very much hoped to keep going until May 1997 when he told me he would become the longest serving contributor of BBC Radio.

I told Bob that my first recollection of him was when, as a member of the "Witney Players" concert party he helped entertain the members of the newly formed Faringdon Fairford, Filkins & Burford Ploughing Society at their inaugural dinner held at The Bull Hotel, Burford, by singing a few songs and telling suitable stories, following the Society's first ploughing match held at William Gammond's Warren Cross Farm, Fairford earlier in the day. The Society was formed with the support of members of the four Farming Clubs incorporated in its title and owed a great deal in its formation to William Gammond, its founder President. Sad to relate not long afterwards William was involved in a farm accident while using a rotovator, causing him to suffer the loss of an arm.

It is good to know that despite one or two minor hiccups, the society stages a well organised and successful annual event. Apart from William Gammond and the various farmers who generously provide sites in the area, year after year I'm sure the Society's existence owes a great deal to past honorary secretaries such as Ray Gilling, Harry Gale, Ivor Cooper and Sid Miller to name but a few.

The event also stages hedge-cutting competitions and I recall with great pleasure that John Jeeves our excellent stockman, always able and willing to take up any kind of farm craftsmanship, won the novice class after a session of practical instruction periods under the tuition of a well known country character named Frank Sykes who hailed from the Banbury area. He was employed by the Regional Agricultural Instruction Committee to teach farm workers the arts of hedge-laying and thatching. I first met him when he came to instruct five or six members of the Faringdon Young Farmers Club in hedge-laying, using the "live stake and binder" technique, as he called it. The hedge, generously provided by Mr John Gilling for us to try our skills with axe and bill-hook upon, was by the side of the lane leading to South Farm, Fernham. This exercise took place in about 1935 and despite the ravages the poor hedgerow must have suffered at the hands of a band of axe-happy Young Farmers it is still alive and well some sixty years since — I do know, as I have visited the site more than once to see the results of our handy work.

Returning for a moment to the ploughing match — it was there that I was impressed by the first demonstration I had seen of a rotary cultivation. It was staged by Messrs Preaters, the Fordson tractor dealers from Swindon and demonstrated by a very enthusiastic salesman, whose face I can remember, but not his name. The Rotavator made by Howards was adapted to fit on to the three point linkage of the Fordson Major tractor and was operated by the power take-off. Following a further demonstration on our heavier land at Baulking I bought the 5 ft wide model. I found it a very useful aid in speeding up the process of make a seed-bed but only if the soil was reasonably dry — to attempt it when wet could be a disaster as the rotating blades seemed to beat the soil particles together rather like the action of wooden butter pats on butter. It was particularly useful for breaking up the top two or three inches of a grass ley a few days before ploughing, enabling excellent burial of the grass and a saving in the amount of cultivation necessary to obtain a decent seed-bed after ploughing. I always thought too this method seemed to get a bit more air into the soil assisting the break down of grass and clover roots resulting in a quicker availability of the fertility built up by the ley.

At this time we were beginning to reap the benefit of having ploughed up almost all of the old permanent pasture leaving about 30 acres, consisting of a handy sized home paddock close to the buildings and other small acreages as access and lie-back areas. Much heavier crops of grass obviously added to the difficulties of haymaking and storage. One or two implement manufacturers were trying to perfect a machine which towed by a tractor would pick hay from the windrow and compress it into medium density bales weighing a little less than half a cwt when the hay was considered fit enough to gather. News soon travelled down the local grape-vine that Ben Mawle, who had just come to farm at Odstone Farm, Ashbury, had invested in such a machine. Known as a Pick-up Baler. It was made by an American firm New Holland and powered by a 32 horse power Wisconsin petrol engine.

It was not long before I took father to see it at work, baling a large field of hay at so much a bale for Messrs W. Packer & Sons in the shadow of White Horse Hill. We were very impressed and the following season we journeyed to Stratford-on-Avon for the purpose of viewing an identical second-hand machine which we eventually bought for £900 together with one ton of baler twine at a further cost of £290 which was more expensive than usual we were told due to a poor hemp harvest in the far east from which plant the twine was made. This type of cordage was replaced by a cheaper man made polythene product a few years later. Our intention was that we should bale our own hay crop and do as much contract work as possible to help pay for the machine. Our original charge was 4d. (old money) per bale and we managed to earn enough to repay the outlay in the first three seasons.

The storage problem was solved by the addition of a triple span dutch barn to accommodate the bales as well as the corn from the arable break, which was in sheaves cut and tied by a binder; combine harvesters were still few and far between and due to the necessity of providing a corn drying and storage set-up to go with it, was only justifiable on larger acreages. The centre span of the barn was useful for many purposes such as short term protection of trailer loads of bales or sheaves, temporary parking of machines and the facility of being able to thresh the corn under cover in the winter. If the corn had been harvested in reasonable condition, it matured and dried out naturally while in the stack, producing a saleable sample straight from the thresher. I still think there was a kind of bloom on the samples of corn harvested in that way rather than there is from the present combine and drier method.

It is a very satisfying feeling to see the corn coming out of the spout and plunge a hand into a three parts full 4 bushel sack to assess how dry the corn was by seeing how far down into the sack you could get your hand. Usually the man in charge of the threshing outfit was able with his wealth of experience to forecast fairly accurately what sort of yield one could anticipate by holding his hand in the stream of corn as it entered the sack. Frank Timms who worked for W.Baylis & Sons was usually not far off the mark when he said either "good" "average" or "poor". It was Frank's job to make sure the threshing machine was operating properly. With the help of his mate, Ernie Rogers who cut the twine, bonding the sheaves and fed the unthreshed straw into the drum, the aim was to provide a fair days threshing as charges were billed by "days" or "half" days. They were a good team and I remember Frank whenever I needed to tie the top of a sack. He would take a bag-tie (twine cut from a corn sheaf) hold the top of corn sack together with his left hand, wrap the twine round the neck of the sack twice, let go and pull both ends of the twine in opposite directions which secured the neck of the sack completely but enabled it to be opened in a flash by merely pulling one end of the string. Frank showed me how to do it many times but somehow I never mastered it and my efforts usually resulted in me having to scoop corn off the floor once the sack was moved and resort to tying a knot which usually required considerable time and patience to undo.

Due mainly to the whims and unreliability of our weather during the grass harvesting season I, like many others, was conscious of the amount of food value that was lost during methods of conservation, especially haymaking. A brief effort on a few acres using wooden tripods failed miserably, due to the amount of labour required and the end product turned out to be very little better than hay made in the traditional way. Silage making was being talked about as the answer and Rex

Paterson, the great thinker and large farmer down in Hampshire had produced the "Paterson Buckrake" for the purpose of carrying fresh cut grass straight from the swath to wedge shaped silage clamps which could be sited close to where it was to be fed during the winter. While I could appreciate it suited his extensive system of out-door dairy farming with 100 cow units dotted about on his rolling flint and clay banks around Hatch Warren in Hampshire, at the time I couldn't think it would be ideal on 167 acres of heavy land at Baulking.

A visit in the mid-thirties to see Clyde Higgs dairying methods at Hatton Rock in Warwickshire where he based the almost entire feeding system for his herd of Ayrshires on dried grass had created a lasting impression on me. So much so that I felt it would be worthwhile some fifteen years later looking into the possibilities and economics of adopting a similar method.

I gathered together all the information I could get and read a great deal in the farming press of what a Mr J.S.Morrey was doing in Shropshire in somewhat the same manner as Clyde Higgs, and attended one or two conferences where he was speaking. I particularly remember him saying that he was achieving maintenance plus three gallons from feeding 20 lbs of dried grass per cow per day. This really set my mind alight and I felt it was a good enough basis from which to delve into what the total requirement would be in terms of capital, running expenses labour reorganisation etc. etc. The latter would be essential as at peak periods the drying plant would be operating 24 hours a day. Incidentally, Mr Morrey was elected a Special Member of the M.M.B. before leaving Shropshire to farm close to Devizes in Wiltshire, where he continued to fulfil a leading role in dairy farming and later became a Joint Master of the South & West Wilts Foxhounds.

Having just started grassland recording in co-operation with the agricultural division of I.C.I. I was able to draw on valuable information from their regional field staff, under the leadership of Mr Frank Henderson, a Senior Development Officer who lived at Tonbridge Wells.

Without exception they were a "great bunch", mostly straight from University. Their advice was backed up by field trials at I.C.I. experimental farms at Jeallots Hill near Maidenhead and Crewkerne in Somerset, the results being disseminated by Peter Flechia and Martin Hutchinson. In conjunction with marketing representatives for the area John Ward and Trevor Lindlar they formed a team very able to give sound advice which covered a much wider area of farming than just persuading people to buy a few tons of fertiliser.

I had visited Jeallots Hill as a member of the Faringdon Young Farmers Club some twenty years earlier. There are three things I particularly remember about that visit. It took place about a fortnight after the Royal Ascot week, which suffered a violent thunderstorm during racing one afternoon. Apart from causing material damage, a bookmaker was tragically killed by lightening while standing under his umbrella. Acres of corn in the area were flattened by torrential rain, including the corn trials at Jeallots Hill. The trial plots measuring about six yards square however revealed an interesting pattern of damage which was plain to see. The corn had gone down in the middle of each plot in a saucer like shape while the edges of the plots dissected by walk-ways for observation were still standing upright, which we were told would lead to experimental plots the following year using varying drill widths as it was thought possible that the strength of stem had something to do with extra light available to the plants on the outsides of the plots.

Another interesting observation concerned grazing management of plots all sown with exactly the same seed mixtures consisting of cocksfoot, ryegrass, timothy, fescue and red and white clover; basically it was known as the Cockle Park mixture. With the use of sheep each plot was grazed at different times which were relevant to the peak growth period of one of the grass or clover species. In this way it was revealed that by so doing it was possible in a short while to finish up with each plot displaying a predominance towards every one of the individual species which composed the original mixture.

By far the most interesting experimental exercise though was the forty acre small holding which was run entirely by a husband and wife team for the purpose of demonstrating that a reasonable living could be made from a small grass acreage well managed and fertilised (of course!) supporting an eighteen cow dairy herd with followers, a couple of breeding sows and a few free-range cocks and hens.

The project was costed meticulously and described in an annual report which indicated that not only did it provide a living in times that were far from affluent, but it illustrated the possibility of perhaps being able to put a foot on the bottom of the farming ladder.

Reverting to the possibility of adopting a grass drying and feeding programme, spurred on by my belief that to extract the moisture from young nutritious grass was as near to perfecting a conservation method as it was possible to get, the only thing that was holding me back was that I was not as entirely convinced that we had got the capital expenditure, and more importantly, the running costs as clear cut and watertight as they appeared to be on paper, despite the help that Alan Harrison of Reading University, Agric Economics Dept and the I.C.I. boys had given to me. In fact we had got so far as deciding a Templewood Drier, Mk. IV was the most suitable for the acreage and size of herd anticipated. However, it was Sam Huthnance our County Agric Officer (Berkshire) who brought me down to earth while on a visit to Church Farm by saying "It all looks O.K. on paper but don't forget every time the fuel tanker comes through your farm gate it will be quite a few quid up your jumper" or words to that effect. — My feet went cold and I chickened out there and then. A decision I'm sure which brought relief to my bank manager when I told him a day or two later, although I think he would have let me have a go — bless him. I was naturally disappointed but tried to console myself by realising the money that would have been spent on fuel oil would buy a considerable tonnage of feeding stuffs and that the alternative was now the time to take a much closer look at silage making and feeding.

So it was in the mid-fifties with valuable help from John Ward backed up by Trevor Lindlar that conferences and farms were visited to gather as much information as possible about techniques which were developing in all the aspects of silage, flail mowers, which cut and blew the grass directly in trailers had arrived on the scene. Trailers equipped with hydraulic tipping gear deposited their loads of grass close to bunkers, usually erected within the confines of existing dutch barns, the floors being concreted and the sides and back constructed of disused wooden railway baulks. Stored in this way silage could be self-fed to stock over a barrier consisting of a metal frame which could be moved forward daily to provide sufficient silage within reach of whatever type of stock was being fed. This was the basic method and equipment we set out with. As knowledge concerning the methods of making and feeding increased, modifications were made to take advantage of work being carried out by various experimental centres under the guidance of people like Dr Jock Murdoch who I remember phoning one day to ask his advice, as we were having difficulty in keeping the temperature in the clamp to a reasonable level. He suggested I borrowed another tractor and trailer to assist in filling the bunker at a much faster rate — sounds a simple answer, but it must have worked, as we managed to win the Silage Cup, presented annually by the Berkshire Grassland Society which was formed in 1962 following a great deal of enthusiastic work by Dick Powell another stalwart of I.C.I. agricultural division. I had the privilege of being a founder member, attending the Society's first meeting which was held at The Ship Hotel, Reading. I also had the honour of being its Chairman the year Dick Powell retired. I think Dr David Allen of Reading University was our Hon.Sec. at the time and we arranged a farewell presentation evening at the Calcutt Lounge Hotel on the Bath road just outside Reading. Our retirement gift was a pint sized pewter tankard with a glass bottom, onto which David had pasted part of a rival advertisement from another well known fertiliser manufacturer which could be read only when the tankard was empty. It was my pleasant task to thank Dick for his past services to the society and hand him the tankard about half full of champagne suggesting that if he was prepared to drink it up straight away there was an important message inside on the bottom of the tankard. It wasn't his first drink that evening but he stuck to the task manfully, exploding with "bubbly" flying everywhere when he reached the point at which he was able to read that "FISONS GROW BETTER CROPS". A short while later with barely a glance in either direction Dick left the hotel car park waving the tankard from his car window while we prayed he would reach home safely at Guildford. A phone call early next morning revealed thankfully all was well and his wife reported Dick as saying "It was a very enjoyable evening".

Most of my time was being spent scheming on the farming front, in order to keep the wolf from the door, and like most parents to provide something towards a reasonable chance for the children to go to boarding school.

In some respects, perhaps because I was an 'only one' I like to think the seven years spent at King Alfreds and Dauntseys helped me a great deal. The rules at both were strict and enforceable by various methods of punishment including up to half-a-dozen stripes across the backside in extreme cases, underlining I am sure what mother and father had taught me about what was right and wrong. I'm also reminded of the words of a dear old friend of the family named Clarkie (J.W.Clarke) who, once said to me just after I left school "You can't play the game boy, unless you have been shown the rules".

He was a very interesting and grand old man whose friendship with my mother began when she was a teenager and helped look after some of his children when they were small and lived next door in London. He was a leather tanner by trade, having been apprenticed to a firm in Bermondsey he formed a partnership with three members of the staff when the tannery came on the market just after the turn of the century. He managed to pay his partners out just before the first world war and struck lucky soon afterwards enabling him to retire in the mid-twenties and pass the business on to two of his sons and building himself a very nice home in Crown Lane at the top of Streatham Common, where he lived with his dear wife who sadly was crippled with arthritis, two unmarried daughters who kept house and looked after them, as well as their bachelor brother. They were all very good to me and I stayed with them many times in my teens and early twenties, when they took me around the sites of London as well as to West End musicals such as "The White Horse Inn", "Mr Cinders", "The Vagabond King" etc.

Clarkie's slice of good luck happened during the very early part of the first world war, when the War Department were seeking tenders from leather tanners for the supply of a large amount of leather of a specified colour to make chin straps for soldiers caps. Apparently, the colour required was proving difficult to achieve, so much so that Clarkie, convinced of the possible rewards, was burning the midnight oil, mixing various dyes with little success when he reached across the marble topped table on which he was working to pick up another bottle and in so doing knocked over three or four other bottles he had been using, the spillage from which mixing with the cocktail already on the table presenting him with a valuable clue which quickly led to the elusive shade of brown he and others had been seeking. It sounds a very small achievement but acceptance of his tender meant a great deal to his business.

I still have a tangible reminder of him and his great kindness to me in the form of a suitcase which he had made up and initialled from a specially selected hide from his tannery and gave to me for my 21st birthday some sixty odd years ago.

Realising that I was approaching the ripe old age of forty and that it was impossible to start again where I had left off on the soccer and rugger scene when war broke out in 1939, I gave up the idea of taking part in any winter sport. Nevertheless,

I was easily persuaded by George Adams and one or two others to renew my subscription to the Faringdon & District Cricket Club, which in those days consisted of about twenty members half of which were farmers, so at hay-making and harvest-time it was sometimes difficult to raise a side, nevertheless as all matches were "friendlies" the results were not quite so important as they are today in League cricket, so we won some and lost some, indulged in a fair bit of leg-pulling and had a lot of fun. The Clubs finances always seemed to be in a precarious state, so much so that on one occasion the Hon. Treasurer who was in fact a local bank manager said at our AGM, we really must make an effort to reduce the overdraft and endeavour to put the finances on a sounder footing. Two or three hours and a few gin and tonics later, it was suggested that each member present should chip in a fiver and we should gamble it all on a horse in the forthcoming Derby at Epsom. It was mutually agreed that Guy Woodin who was considered to be the most experienced member among us in such matters should be entrusted with selection of the most likely winner. Alas his confidence in a horse named Marengo backed each way at 100/6 to get us out of trouble, despite a brave effort failed to finish in the first three, with the result that about a dozen of us were a fiver poorer and the club finances remained the same. They were however, restored to an acceptable level by a bumper sweepstake on the "Stewards Cup" run annually at glorious Goodwood in July and a successful Club Ball held the following winter.

The sweepstake promoted by our Hon.Sec and myself with the willing support of most members became an accepted and successful fund raising effort for quite a few following seasons.

I remember a few anecdotes connected with the Clubs' existence during that period. Two of our most enjoyable games were Sunday home and away fixtures with Evesham. They were always a bit too good for us so there were many suggestions put forward as to how we could improve our chances of winning while some of us were rolling the pitch on

Friday evening in readiness for their visit the following Sunday.

We already had a very useful opening bowler named Stewart who was articled at the time to Hobbs & Chambers our local auctioneers but we desperately needed someone a bit quick to back him up at the other end. A suggestion was made that we imported a member from the Cirencester C.C. who had frightened most of us out a fortnight earlier in Earl Bathursts' Park at Cirencester. A phone call indicated that their Club had no Sunday fixtures so he would be very pleased to turn out for us at Faringdon. That it was perhaps not the sportiest thing to do but as there was nothing but the glory of winning at stake we would keep our friends' appearance as low key as possible and appease ourselves with the knowledge that it would be a better game instead of a foregone conclusion as in the past. It so happened that Evesham won the toss. Deciding to bat, their opening batsmen, Sam Johns and Mike Collins arrived at the wicket, whereupon my neighbour, Maurice Reade who had just given up playing for us and become our 'unofficial' umpire stepped up to the stumps to inquire of Mike Collins what guard he would like, to which Mike replied, after taking note of our Cirencester friend marking out a long "run up" and where I was positioned as wicket-keeper that he would take "middle and leg from where the bowler bowls" to which Maurice replied "I haven't seen him bowl yet". So much for our little scheme to achieve at least one result in our favour without disclosing we had enlisted outside help — we still lost anyway!

Our away game with Evesham was always a very enjoyable day out, most of us meeting up for lunch at "The Crown" in the centre of the town before arriving at the picturesque ground situated by the side of the River Avon which occasionally swelled over its banks, necessitating the pavilion which also accommodated tennis and hockey clubs, to be built on stilts, therefore requiring quite a large flight of steps for access. I mention the latter as one of our members was rather prone to 'shooting a line'. He had collected quite a selection of well known cricket caps. How he got them we never knew, but to enhance his ego he would appear at the crease or in the field adorned as a member of clubs such as Free Foresters, Eton Ramblers, MCC and many others. Someone "nicked" a few of them from his bag in the changing room and distributed them surreptitiously to most of the Evesham players who donned them while coming down the steps to take the field of play, having won the toss and put us in to bat. It was quite a few minutes before our member, sitting on the balcony realised someone was "taking the Mickey".

It was on the same Sunday that an incident occurred in which I was involved and have never been allowed to forget by a few of my contemporaries. My dear wife and her sister Em, with our daughter Pam, aged about 2 months had dropped me off in Evesham at lunchtime and journeyed on to see their cousins, the Hillman family who farmed and ran a Stud at Stockwood, close to Feckenham. It was arranged that they would make their own way home, while someone would give me a lift back to Faringdon and Maurice Reade would see me to Baulking from there. Not surprisingly the post match conviviality became somewhat lengthy and I recall some sort of effort being made to beat the number of empty beer bottles left behind by R.A.F. Rissington a fortnight earlier. However, it was late evening when we left and later still when we passed a couple of cars parked on the side of the road half way up Fish Hill on the climb out of Broadway towards Stow-on-the-Wold. It was obvious that they were in some sort of minor bother, so being in a happy frame of mind we gave them a cheer as we went by. On arrival in Faringdon Market Place it was discovered that Dick Taylor's little green Morris Van had a puncture — Dick was one of our regular players being a very good all-rounder. He farmed at Stadhampton, more than 20 miles away, loved his cricket and afterward sensibly stuck to a maximum of a couple of halves of shandy. It was now well past midnight and Dick was unable to find a wheel jack anywhere in his vehicle so eight of us — three each side, one at the front and back — lifted it clear of the ground and held it aloft while Dick changed the wheel. Having waved Dick on his way home, a playing member of our party who was stationed at the Marine's camp at Butts Close on the outskirts of Faringdon suggested our exertions needed a little revival treatment so took us to the sergeants mess for an hour or so of snooker and more liquid refreshment after which we all left to wend our various ways home. When Maurice Reade dropped me off at Church Farm I was taken aback to see lights on all over the place and I soon learned from Ann and her sister that they were the unfortunate people with a puncture we had shouted to half way up Fish Hill, and that a couple of kind gentlemen had stopped to help. That they had only just arrived home was due to the wheel-jack breaking and the necessity to improvise a lifting operation by the use of some stones from a cotswold wall and a piece of rail pulled from a nearby hedge, which was cleverly thought out and assembled by their kind helpers.   As far as I was concerned, I have never been able to live down the fact that my voice had been recognised as we passed them by and had caused Ann to say "That was my husband" or words to that effect!

Another fixture I always enjoyed was with the South Oxfordshire Farmers whose ground was on the village green at Marsh Baldon. It was unique in more ways than one as the pavilion was on one side of the public highway and the square on the other which meant that batsmen crossing the road to reach the wicket and players fielding close to the road were in danger of being run-over. The location was set in a truly rural scene consisting of almost a square of farmsteads and cottages skirting the sizeable village green, with the parish church situated adjacent to one corner of it. On those Sunday evenings when a match was taking place a strange but short charming little episode took place which I witnessed a few times and it never failed to move me. At around ten minutes to six the church bell would ring out calling the faithful to Evensong. This was the signal for a cottage garden gate at the far end of the village green to open, revealing an elderly gentleman in his Sunday suit starting his slow but purposeful walk to church along the public footpath he had trodden for years and which took him within a few yards of the wickets, where he paused for a moment to touch his cap and say "Good evening, gentlemen" before going on his way, enabling play to resume once more. If the match hadn't finished when he came out of church, the procedure was reversed, except that he said "Goodnight, gentlemen".

At that time most of the groundsman's work was done by a great character and sportsman, named Ron Wells whose farm was adjacent to the Green. He was always good fun, very good at telling a story, in his own inimitable style and a great leg-puller. Aided and abetted by his great friend Ted Gent he was fond of getting non-smokers to accept a cigarette when they were being offered around usually in a pub bar and get them to stealthily pass them on to him to put in his own packet to offer back to you, so that you were smoking your own — in this way his packet always appeared to fill up, instead of emptying as the evening progressed.

A friend of mine Tim Weaving who once played for the South Oxfordshire Farmers told me that following a Saturday evening's rook shooting he had left half a dozen young rooks in the boot of his car by mistake — after Sundays match he pushed a rook into each of Ron's cricket boots while in the changing room, only to receive them by parcel post a fortnight later with a note saying "Dear Tim I think these are yours — sorry not to have returned them before this" Yours ever Ron.

Doug Surman who with his brother Rusty also played for the S.O.F's told me a story recently, concerning Ron. Apparently, he was attending a funeral of one of his cricket contemporaries and with one or two others, had propped himself against the churchyard wall close to the gate. Following the end of the service, as they watched people leaving, he noticed two very elderly cricketers who he once played with shuffling slowly towards the gate with the help of a couple of sticks and Ron was heard to remark "Look at those two poor ol' so and so's 'tis hardly worth them going home, is it?"

He often rang me up in the evening time to have a chat and a laugh but sadly he died recently in his 80th year and I miss his earthy remarks and dry sense of humour. One of the last times we spoke, I suggested we met with our wives and had a meal out one evening — to which he quickly replied — "Not bloody likely they charge £5 a bottle for drawing the cork these days and I'm not having any of that". On another occasion knowing he was suffering with arthritic knees I asked how he was and he said "Not very well, I get up in the morning, sit in my chair and wait to go to bed at night."

I'm sure this unique sense of humour and quick wit did much to help relieve some of the pain he suffered. I can well imagine the enjoyment he derived from an incident related to me by a mutual friend, Jimmy White. It concerned a stranger passing through the village, stopping his car to inquire of Ron if he knew the time? "Of course I do," said Ron in his usual slow, polite manner. "Tis winter-time and ain't it ruddy cold."

There seem to be so many other things I could say about the fun of being involved in local cricket. The friendly and wonderful people I have met on and off the pitch during an innings of thirty odd years have given me many names to remember and a great deal to treasure.

TEN
Wireless & Telly

It was on the 8th of May 1953 that Robert Pocock of the BBC came to see me at Church Farm, following a suggestion made to him by John Betjeman that I might fit into a programme he was arranging in a long running series of broadcasts on the B.B.C's Home Service entitled "Country Magazine". It was to be transmitted live from Studio 3A at Broadcasting House on Whit Sunday May 24th 1953. Introduced by Ralph Wightman and produced by Francis Dillon. It concerned various aspects of life in the county of Berkshire as seen through the eyes of a few people living within its boundaries. A written description of my modest farming activities was accepted and I was instructed to report to Rothwell House just off Gt. Portland Street for rehearsals the day before. As I approached the main entrance the commissioner welcomed me with a smile and the words "Country Magazine Sir? — Take the lift to the third floor". The following 36 hours were very interesting indeed . There were seven of us in the studio to speak for ourselves. Mrs Sally Hogg, a Postmistress from Yattenden, Geoff Whatley an iron founder from Bucklebury, Joe Lammin, a stableman at Lambourn, Fred Ilott who worked for a well known seed firm near Reading. Mr Gaythorne Hardy, a naturalist also from Bucklebury, Mr Culley a water bailiff who looked after a stretch of about five and a half miles on the rivers Kennet and Dun and then there was me doing my best to keep my end up on behalf of those struggling to wrest a living from the soil in the Vale of the White Horse at the foot of the Berkshire Downs.

The whole thing was quite an experience from start to finish, following introductions and a period of getting to know one another we were handed edited scripts made up from our own originals and put together in a conversational manner for us to read through to ourselves. We were then taken into one of the fully equipped broadcasting studios for the first "read through", using a "dead" microphone but treating the exercise as if it was going out "live".

We were all introduced into the programme individually by Ralph Wightman. Being a Dorset farmer as well as a well known broadcaster, his cue to me was "being farmers we have something in common but I don't think I have been to your farm", to which knowing full well he hadn't been within 20 miles of it, I squared my shoulders, moved closer to the microphone, suspended by springs in the middle of metal or bakelite ring with B.B.C. mounted above it, and made my reply "No, I don't think you have Ralph" in what I thought was an attempt to be loud and clear. Whereupon the producer who asked us to call him Jack, shouted "stop" descended by the stairway from his sound proof, glass fronted observation point, put his hand on my shoulder looked me straight in the eye and quietly in front of everyone inquired if I had ever seen a microphone before. He went on to explain what marvellous things they were and when switched on and spoken into, it was possible for people all over the world to hear what you were saying and there is "No need to shout like that or you'll break every "mike" we've got in Broadcasting House". I immediately shrank to half my normal size and so in turn did all the others as he systematically levelled us all and then lifted us to the point when we all thought our name was Richard Dimbleby. It would take too long to record the various ways by which he achieved his purpose but I recall how he dealt with the only lady member among us. As Ralph Wightman gave her the cue for her introduction to the programme; she really got going. Descending once more to the studio floor level Jack Dillon put his arm around her saying "Darling, this programme is timed to last half-an-hour — at your rate of babbling the whole thing could well be over in 10 minutes".

There are quite a few things I remember about that programme such as Ralph Wightman asking me during our chat about farming in North Berks if there was difficulty in keeping labour on the land or if there was a drift towards the larger towns and the light industry that seemed to be starting to grow up around them. I answered by saying that farming was a seven days a week job and until someone was able to breed a cow that only milks a five day a week there will always be a bit of a problem. Never the less I felt the real farm worker with his heart and soul in mother earth, provided he was given a wage comparable with industrial workers would never wish to leave the land. I remember going on to recite a story of this brand of loyalty which was illustrated by the old carter who was given the sack for running his wagon into the gate post when returning to the farm late one evening after making too many calls on the way home, following the delivery of twenty quarters of wheat to the local corn merchant. Next morning, however, when the farmer was doing his rounds on his old cob, there he was turnip hoeing. "Didn't I tell you never to come back here any more?" "Thats right, sir, so you did" answered the old boy, "but if you don't know when you've got a good man, I knows when I got a good boss — and I ain't a-goin".

I recollect this little story producing quite a volume of spontaneous laughter from the Wynford Reynolds Sextet whose musical rendering of "The Painful Plough" played the programme in and out as well as providing accompaniment to Marjorie Westbury who sang what she described as a famous old Berkshire song call "The Berkshire Tragedy" — It may have been famous but in fact neither of us had ever heard it before.

It was interesting to meet Miss Westbury as she was currently playing Paul Temple's assistant in the Radio series of detective programmes. Her voice both speaking and singing was most appealing but portrayed someone totally different to the person

I had imagined when listening to her on the radio at home. Also, by a sheer coincidence we were all treated to a grandstand view of some of the famous people who were arriving in the reception area of Broadcasting House to take part in the Saturday evening production of "In Town To-night", one of whom happened to be the film star Katherine Hepburn. Despite strong-armed assistance from the police and B.B.C. ushers it took ages for her to get through the hundreds of fans and autograph hunters who packed the roads and pavement outside. Having seen her at close quarters I'm not surprised.

It wasn't until some while after this programme that I was invited to what was supposed to be an occasional chatty account of what was happening in and around the Vale of the White Horse. This was for the B.B.C.'s programme " Farming Today" hosted usually by David Butler a farmer and broadcaster from Hampshire. He would ring me up at 6.10a.m. on a Saturday morning from the studio in Bristol for a "live" chat to slot into the programme for a few minutes between matters of current concern on the farming front which were disseminated by experts in their own particular agricultural spheres. I sometimes felt it was a bit too early in the day to be at one's best and often wondered if a couple of "gin and gingers" might have helped — but never got round to putting the idea to the test — perhaps from fear that I might overdo the dose.

This occasional involvement led to taking part in a broadcast known as "On Your Farm" which was from the kitchen at Church Farm, complete with the sound effects of sizzling eggs and bacon. This happened one early morning in 1966. It was produced and presented by Tony Parkin, using the kitchen as the focal point of the roving microphone to each of the eight dairy farms in the village for a brief chat on what was actually going on; the problems, if there were any and of course hopes for to-morrow. These brief interviews were conducted by David Butler and a well known Norfolk farmer David Richardson who, as well as broadcasting, currently writes a weekly full page article for "The Farmers Weekly". The interviews were transmitted to the B.B.C. receiving van at Church Farm, from which they were fed live into the national radio network. While the two Land Rovers were travelling from farm to farm the programme returned to Tony Parkin in the kitchen where he maintained continuity by talking to employees and members of the family, with the exception of our daughter Pam who was absent, having been selected as one of half-a-dozen members of Young Farmers' Clubs to represent Gt Britain on a nine month trip to Australia. The group sailed to and fro as first class passengers and guests of the company on P & O Ship SS Iberia and they lived with Young Farmer families during their travels around Australia.

Apart from radio engineers, producer's assistants and the like there were also a few local friends around the breakfast table; one of whom was Colin Nash, a great friend and beloved Master & Huntsman of the Old Berks Hunt. It was arranged that Colin's arrival outside should be accompanied by the sound of his hunting horn and answered by a "holla" from within, the result being that Colin delayed blowing his horn until he was in the hall and coupled with my holla from the kitchen sound engineers were almost forced to leave their posts and feared that the equipment may have suffered irreparable damage. However, all was well. That evening I had a telephone call from the hospital bed of Dick Petit-Mills, he was a lovely chap and suffered a great deal — having an artificial leg. He farmed at Brimpton near Newbury and did a great deal for the N.F.U. being a past County Chairman. He told me that he was feeling a bit low, lying there with headphones on fairly early in the morning when the programme came on. He said he was beginning to chirp up a bit when all of a sudden the sound of the horn accompanied by the "holla" almost had him jumping from his bed and he wanted us to know that it had done him more good than the whole of the medication he had received during the past fortnight. Bless him — sadly he is no longer with us. Somehow I couldn't help thinking there is something in the old saying "It's an ill wind that doesn't blow some good".

It wasn't until 1974 that I was fortunate to have the experience of witnessing the production of a 30 minute programme for a T.V. series being produced by Roger Mills for the BBC entitled "Inside Story". It was a series covering the inside stories of a very wide range of subjects — from Pop Groups to the production of a pint of cow's milk. It was to tell the story of the latter that Roger first came to see me, for the purpose of discussing what he had in mind; which was basically to start with the birth of a calf and to follow the results of the event in terms of the mother's life, her off spring and the milk inevitably produced. Having selected a cow named Celia from half a dozen due to calve within the time specified by Roger for the purpose of being able to have the film crew on site in readiness. Unfortunately, Celia had her own ideas as to when she would give birth and despite predictions including one from the local vet who disappointed Roger Mills by saying she would calve when she was ready, probably in about 48 hours but due to the array of arc lights set up in the calving box it could be much longer. Due to this forecast and the fact that Roger's financial budget was coming under severe pressure he decided after two nights vigil everyone should take a night off and report back at Church Farm at 8 a.m. the following morning. Sure enough a glimpse into the box before starting milking at 5.30.am. revealed that Celia had calved, cleansed and was lying down chewing the cud, pausing momentarily to look at the calf by her side and then at me with almost a wink in her eye. But there was more to come. Roger had chosen the title of the film to be "To Celia a Son" and had already shot parts of it under that guise — but a lift of one of the calf's hind legs revealed it should have been "To Celia a Daughter".

All this meant that when Roger and the camera crew arrived a great deal of banter ensued followed by filming a simulation that evening after dark of what actually took place, in which a bedroom light was switched on and a moment later I was seen emerging from the back door with a lantern to make my way to the calving box where I discovered the calf had arrived safely and all was well, before returning to bed.

The actual gender of the calf was concealed by being particularly careful to film a frontal view. This was especially so, when it was ten days old and being auctioned at Swindon Cattle Market. The whole programme necessitated quite a few visits to Church Farm and other relevant venues such as dairies and markets, taking a few months to produce. On completion our small staff of three John & Leslie Jeeves and Charlie Scriven with son Bill and myself were invited by Roger Mills to a preview at the B.B.C. Television Centre in London which was a very interesting day indeed. As well as seeing "To Celia a Son" we were taken on a tour of productions in progress such as "Top of the Pops", "Dixon of Dock Green" and the set used for filming "Steptoe & Son". We were also lunched in the T.V. Cafeteria where we recognised a few of the current stars of the day — "Z Cars" in particular.

Another programme which I found very interesting and had a little to do with was named "Cusden on Location". It was first shown on B.B.C. South appearing later on B.B.C.1 in March 1980 and was produced by John Cox from Southampton. I first met him when he had something to do with a short film connected with our old village horse and hero Baulking Green. He rang me up to ask if I would take him round our neck of the woods for the purpose of showing him a few old farmhouses and cottages that might fit in to a series of three programmes he was producing which featured an artist named Richard Cusden who was to set up his easel in or around old homes for the purpose of sketching and describing what he saw, its architectural and possible historical background relating to the people who had lived in them during the last few centuries. Some of the locations chosen included the farmhouses of Manor and Church farms as well as the old thatched cottage close to Baulking Church which, when father came to Church Farm housed three separate farm workers families. It was converted into a two family dwelling with the assistance of a grant provided under the Housing (Rural Workers) Act in the late twenties which fixed the maximum rent chargeable for each cottage at three shillings and five pence per week for a period of 20 years. Another inclusion in the programme was a very small but fascinating thatched cottage in the village of Westcot near Kingston Lisle.

I hope the recount of my occasional experiences in this particular field spread over some thirty years have not been too boring, but I think a few glimpses behind the scenes has at times broadened the angle from which I view television or listen to the radio.

ELEVEN
The Sixties

Any attempt to write my memoirs would be significantly incomplete if I did not mention the tragedy of my father's death in a tractor accident in an arable field on Barrowbush Hill near Fernham on August 29th 1961.

I trust I can be forgiven for being brief, but even now I am reminded quite often for a fleeting moment of some aspect of the sudden drama of which I was so much a part.

It still seems so unreal that seconds before it happened we lit cigarettes together and father remarked: "Wonder who bought Long Lane this afternoon?" (Long Lane was a plot of approximately 90 acres of farming land very close to Barrowbush, between Rosey Brook and Fernham which at one time belonged to my grandfather and was being auctioned that afternoon). It was inevitable going through the tragedy in my mind many times afterwards that those two small words "if only" kept appearing. "If only" one of the two trailers we had loaded with straw bales had not suffered a punctured tyre then father would have taken it to Baulking instead of riding home with me when a bale fell off the front of our load knocking him off the tractor.

We know it is almost impossible to prevent such eventualities but accidents are usually due to a chain of circumstances so to anyone reading this I beseech you to *take as much care as is humanly possible* when using farm machinery of any kind especially the huge pieces of modern equipment that are part of today's farming scene.

It was sometime before we as a family began to recover

from the shock. Daughter Pam, and son Bill who was with me when it happened, were in their early teens. Their very presence was a tremendous consolation and encouragement to Ann and myself so with the help and support of many friends we gradually got back on the road to living but never forgetting, accepting in our innermost hearts that however tragic and grievous such a loss can be life has to go on.

Looking back I am very conscious of the great help it was to me to be invited to take father's place on the Old Berks Hunt Point to Point Committee. Father had been Hon.Sec. for the previous fourteen years, the last nine of which being associated with the course layed-out by Mr Christopher Loyd on part of his estate at Lockinge in 1953, a venue which has been graced by the presence of the Queen Mother, and he has hosted the fixture ever since. Its extensive natural hillside grandstand, providing such excellent viewing, continues to attract huge crowds every Easter Monday.

It was in 1957 that Queen Elizabeth, the Queen Mother accompanied by Princess Margaret came to Lockinge to see Her Majesty's horse Gipsy Love beaten into second place in the Open Race, by a very good horse named Grey Spot belonging to an erstwhile stalwart of the O.B.H. Victor Arkell. Despite heavy showers of rain during the afternoon the Chairman of the Point-to-Point Committee Bob Pike and my father as Hon. Secretary had the great honour of being presented to Her Majesty by Mr Christopher Loyd. Standing close by I remember seeing The Queen Mother put father at ease by saying that she understood he had hunted with the Old Berks for sixty years and she hoped there would be many more to come.

To do justice to the sport of Point-to-Pointing and the fun it has given me ever since attending my first meeting a mile from here on Baulking Hill in 1921 when eight years old, would probably fill another book. I can only say during the seventy-six years that have followed I have had the privilege of meeting scores of lovely people connected with the sport from those who organised it and made it all possible to those who like me in the main just stood and watched on courses as far apart as Garthorpe in Leicestershire, Llantwit Major, and the Golden Valley in South Wales. As well as many others in Herefordshire and the West Country such as Bratton Down and Munsy Hill Gate on Exmoor, not forgetting the ever popular meeting staged by the Torrington Farmers at Umberleigh with its distinct "end of term" atmosphere.

In the late eighties I remember two conflicting reports appeared simultaneously in The Daily Telegraph and The Sporting Life as to the venues of a short list of fixtures for a Saturday close to the end of the season,, Not taking the trouble to verify, we thought it would be nice to explore new territory so we set off to Llantarnham as per recommendation in "The Life". The outcome of our exploit if nothing else, I felt was worthy of being recorded in poor poetic form, which my friend Michael Williams point-to-point correspondent of the said newspaper acknowledged in the manner I hoped for and the leg-pull was put to rest in the secretary's tent at Lockinge the following season to the accompaniment of a couple of tots of Gin and Ginger.

My humble effort to record the occasion in verse went thus:—

>   Early one morn, when the sun began to shine
>   End of season's Point to Pointing very much in mind
>   Fixtures few and the going hard
>   Long priced winners difficult to find.
>
>   We searched among the racing news
>   In Saturday's 'Sporting Life'
>   To see where Michael Williams suggested we might go
>   And to check his good advice.
>   Sure enough there it was
>   Plain for all to see
>   A place named Llantarnham, where t'was said
>   The Ystrad fixture was to be.

We gathered all the things together
That 'Pointing' followers like to take
Field glasses, sandwiches, supplies of Gin & Ginger wine
And for tea — a lump of mother's currant cake.

We packed the lot into the boot
Phoned a friend — told her we were going far
Would she like to come with us?
There was room within the car.

Away we went journeying on with speed
Along the motorway — across the River Severn
Past many sets of rugger posts
Echo-ing "Bread of Heaven".

At last — four miles north of Newport
T'was here the races should have been
But alas! despite many enquiries
Neither tent, fence, horse or rider could be seen.

From what a friendly female native told us,
Others too that day had suffered desperation
While "it was in Llantarnham that we were"
Some thirty miles away, near Cowbridge, should have
                                        been our destination.

However — Fortified by three halves of best Welsh bitter,
And the sound of well mean't instructions from other
                                            native lips
We made the third race at Llantwit Major
And backed two of Michaels' winning tips.

Alas — another season has come and gone
Sheep and cattle graze where bookies used to shout
But February will bring the fun again
And FUN is surely what 'tis all about.

R.L.

Another story might be worth relating concerning what must have been almost the last time the Old Berks fixture was held at Middleleaze Farm Coleshill by kind permission of that great character Darby Butt. It must have been about 1949/50 and as usual the Butt household was open all day providing unbelievable hospitality to one and all. With racing completed for the day my father who was Hon. Secretary had enlisted the help of a friend Jimmy James who was the current head cashier at a well known bank's branch at Faringdon, to help him as unofficial treasurer.

The two of them having tidied things up had adjourned to Darby Butt's sitting room from where I was instructed to pick them up at around 8pm — having myself been home to help finish off evening milking. Arrangements had been made with the bank manager for Jimmy to ring the night bell at approx 9pm when he would take the day's cash takings into safe keeping. A brief survey on my arrival back at Darby's indicated that this was going to be a lot easier said than done — and it was sometime before I could even persuade anyone that Mr Fitz, Jimmy's boss would be waiting to deal with the money and go to bed. Mention of money did however add some seriousness to the situation as no one could remember where they had put it. One can laugh about it afterwards but trying to conduct a search by a number of well meaning but mostly pie-eyed people had to be seen to be believed.

Fortunately Darby's sisters Camilla and Clare were able to provide the sobriety necessary to discover that Jimmy and father had hidden the leather bag containing between £700 — £800 behind the hat-stand in the hall, and had carefully camouflaged its whereabouts with a few overcoats. Following a further "one for the road" I managed to get them both towards the car in the stable yard close to the back door when, Jimmy expressed the wish to "water the pony" which Darby suggested might be done in the nearest stable across the yard. Whereupon Jimmy mumbled something about not being able to wait and proceeded to relieve himself on the spot, causing Darby to shout "Not there" at the same time giving him a quick thump in the middle of the back, which resulted in the immediate dislodging of Jimmy's set of false teeth, adding further complications, which I won't go into, to an already unfortunate state of affairs and further delay to our already belated departure for Faringdon.

However our arrival in front of the bank, left only one final problem to be overcome. It concerned Jimmy's declared intent to hand the money to his boss in person and the difficulty he might have in trying to walk up the three stone steps to the bank door. I can only say after quite a few tries he was only able to carry out his wish with my assistance from behind from where I was also able to observe his boss's silent reaction to what was happening. Suffice to say Jimmy remained a well respected member of the bank staff until he reached retirement age and fulfilled honorary secretarial duties on behalf of the Faringdon and District Cricket Club, the Faringdon Rugger Club and other local organisations. We attended many matches at Twickenham together and I remember him, not only for his friendship but for the great sense of humour that went with it. He was once heard to mutter quietly "Does he think we are proprietors of a public convenience as well as bankers". to a fellow member of staff whose help he had enlisted to count an enormous amount of pennies, required as change by a customer organising a village fete.

My interest as a follower of sport "between the flags" and the fact that I had been assisting father a little with some of his secretarial duties had the effect of providing me with interest that contributed much towards the period during which, without being aware I was undoubtedly struggling to see and think straight once more, following father's tragic accident.

It was also at this time that the legend of our old village hero "Baulking Green" the future champion hunter-chaser was beginning to unfold. In fact following his first appearance at the V.W.H. meeting at Barbury Castle on March 29th 1961, when he was beaten a neck in the Adjacent Hunts Maiden he appeared five days later in the O.B.H. Hunt Race at Lockinge.

Ridden by one of our best loved and remembered M.F.H.'s Colin Nash, he was a little surprisingly (I thought!) beaten four lengths by John Mason's Prior Approval, another useful and consistent performer, with whom John had a lot of fun, winning quite a few races with Rosemary Lomax aboard. Colin was heard to remark after the race that if he had been as fit as the horse the result may have been different — but who knows — that's what it's all about.

I am reminded of a somewhat similar story when another Old Berks M.F.H. in the person of Mr G.S.L.Whitelaw, a particularly heavily built racehorse trainer from Antwick Manor, Letcombe Regis, who in the mid thirties dismounted in the winners enclosure at the O.B.H. meeting, over looking Step Farm, on the outskirts of Faringdon, to hear someone close by voice the opinion that his mount named Milk Punch must be a good horse to have beaten "Plodder" ridden and owned by a well known farming character Bob Pike of Lyford. A comment which persuaded the Master to comment; "To have carried three stone overweight and three parts of a bottle of whiskey over three and a quarter miles and win like that, he's not only a good horse, he's a ***bloody*** good horse".

Mention of Bob Pike recalls a couple of stories — the first of which he told me in his car while taking me to play cricket at Cirencester Park when I was about seventeen years old. Apparently on two previous occasions a bunch of Bob Pike's in-calf heifers had broken through a neighbour's hedge and caused damage to a field of grass intended for hay. The neighbour had written describing the damage and asking for financial compensation which Bob Pike considered a very unneighbourly thing to do; he told me he had ignored the matter initially, but had kept the letter safely, and when the incident had occurred in reverse he had taken the letter from his bureau, replaced his neighbour's address and the date of the original incident with his own, updated the letter, crossed out his neighbour's signature, signed the letter himself and two years later posted it back to him, inferring that he knew if he kept the letter long enough he would be able to use it. Such matters are of course usually dealt with by insurance companies these days and in most cases their settlement contributes something towards good neighbourliness.

Bob Pike was a very likeable man. His family, having lived and farmed at Lyford Manor for two or three generations, had without doubt come to earn a great deal of respect and this was a characteristic that Bob Pike obviously felt was something to be preserved among those with whom he came into contact for the benefit of all.

Little wonder that when he thought someone had fallen short by addressing him as Bob he wanted him to know that he was "Mr Pike to his acquaintances, Bob to his friends and Sir to buggers like you". Maybe in this present day and age, there is something in what he said in relationship to some people's lack of respect for others, or their property.

TWELVE
Ol' Baulking

It wasn't until Ol' Baulking, as many of us called him, had run his last race on 8 March 1969 at Newbury when he sadly broke down at the age of 16 yrs that I gave the thought of writing a book about him any serious thought.

Ann and I were fortunate enough to be helping celebrate John Carter's wedding to Judy McLaren. The reception was held at The Rose Revived on the banks of the river Thames at New Bridge. After a few glasses of that which gladdens the heart of men and women and a long chat with the late Phil Wentworth about all the fun we had following the old horse during the nine years he dominated the Hunter Chasing scene on racecourses as far apart as Ayr in Scotland, Wincanton in the West Country as well as Newbury and the headquarters of National Hunt racing at Cheltenham, where he won the United Hunts Cup four times when that race was featured as part of the annual Festival Meeting fixture in March, Phil Wentworth ended our reminiscing by saying "someone ought to write it all down," and suggested I had a go. By the time John and Judy had left for their honeymoon and I had consumed another glass or two of champagne I really felt I could write **anything** down.

However, in the cold light of day, that brand of optimism quickly faded until it was revived a few weeks later by a note from Phil enquiring how the book was progressing. Somehow this galvanised me into action. I equipped myself with pad of foolscap and a pen and I started off into a task about which I knew very, very little, so little that after writing the first few hundred words I said to Ann, "I've a feeling I should put a comma in somewhere".

The book's compilation, completion and publication over a period of two years was achieved with the help and encouragement of many people and gave me an immense feeling of humble satisfaction. I treasure the letters I received from friends known and unknown. There were also people who called to see me and asked to be taken to Manor Farm to see the old horse and meet the breeder and owner — Jim Reade.

One such visitor knocking on my door early one Sunday afternoon just after I had planted my feet on the mantelpiece prior to a little 'shut-eye' before afternoon milking, to say he had read the book 12,000 miles away in the north island of New Zealand, having emigrated there as a young man between the wars. Apparently, he was visiting a friend out there who said he had been sent a copy of a book about a horse named after a village in the old country and some of the people connected with his success as a hunter chaser; adding that he thought he might find it interesting. Looking at the title he exclaimed "Baulking Green, I can't believe it, I was born at a village named Baulking". Flicking quickly over a few pages he went on to say he remembered going to the village school with some of the people mentioned. He then went on to tell me that having taken the book home and read it he made up his mind to fly over with his wife and have one more look at the old country in which he was born. In fact his father worked at Baulking Grange, on the outskirts of the village, prior to saving up enough money to enable him to secure the tenancy of a County Council small holding at Charney Bassett soon after the First World War. He reminded me of the hospitable nature of Sid Reade, Jim's father and his particular love of fox hunting and those connected with the sport. He went on to tell me that his own father, when he could find the time, enjoyed an hour or two following hounds on the old cob which he also used to take the milk from his small herd of shorthorns every day to the G.W.R. station at Wantage Road. On one occasion

Rosey Brook needed to be jumped but, despite a brave attempt, claimed both horse and rider the latter getting off with a 'ducking' but sadly his faithful old cob broke his back and had to be 'put down'. after being pulled from the brook by a trace-horse quickly sent to the scene by Sid Reade.

Apart from being emotionally upset he said his father was at a loss to know how he could manage without his cob in the immediate future. His worry in this respect was greatly relieved by Sid Reade saying — I've got a partially broken four year old 'cobby-type' at Manor Farm you can have which could suit you. "But what about paying for it Mr Reade?" was answered by "That's alright — you call in at the farm and take him home with you and if he'll do the job for you, you can pay me when you can afford to". I was told that is exactly what he did. I can only comment "what kindness and WHAT TRUST" and marvel at the circumstances which came together to persuade Arthur Bayliss and his wife from 102 Mannawapou Road, Hawava, New Zealand to travel twelve thousand miles to visit his birth place and call on me to write "Thanks for the book" in our visitor's book on 23rd April 1972.

There were other instances such as buying an evening paper outside Chester Cathedral the eve of Red Rum's first Grand National win when he beat poor ol' Dick Pitman on Crisp, who was 'out' on his legs, in a desperate finish. I enquired jokingly of the fairly elderly newsvendor if his paper contained to-morrow's winners "Ah, an' the losers as well" he chuckled. Holding his hand out for the few coppers he continued "What's goin a win it d'you think then Guv?" I confided in him that we had come quite a way and hoped Grey Sombrero would oblige. This prompted him to ask where we actually lived. "Oh, a little village, down in Berkshire named Baulking Green, don't suppose your've ever heard of it" says I. "If 'tis anything to do with that ol' 'oss, I certainly do guv — corr — 'e was a good-un 'ee was, I an my mate Charlie won quite a few quid on 'ee". All this meant we stayed quite a bit longer than was necessary just to buy a paper but enjoyed meeting the old boy and realising once again that perhaps the world is a much smaller place than we had originally thought. I wouldn't be surprised if he was a quid or two lighter the following day as poor Grey Sombrero broke his shoulder on landing at the 'The Chair' when in second place behind Pendil and very sadly had to be destroyed. He belonged to Frank Caudwell, one of the foremost stalwarts of farming and hunting in the Old Berks country and following early success between the flags went on when trained by David Gandolfo to win the prestigious Whitbread Gold Cup at Sandown in 1972.

It was the only time Ann and I went to Aintree on Grand National Day. We walked the course and like everyone who has done so, were amazed at the awesome appearance of some of the fences, concluding that this really was the greatest steeplechase in the world and to complete the 4.5 miles, let alone win required the utmost skills of horse, trainer and rider alike. Sadly our day was marred by the loss of poor Grey Sombrero and I shall never forget the sight of Frank Caudwell's wife Doreen, a lone figure walking down the course to 'The Chair' fence, despite pleadings from others not to do so, for the purpose of pulling a corner of the tarpaulin sheet back for a pat on the neck for the old fella and no doubt to say 'thanks and farewell'.

It is true to say that approaching the age of sixty I never dreamt I would write a book about anything. It is also true to say that having done it I never dreamt, sought and certainly didn't expect the kind expressions contained in the sporting press reviews which followed the books publication. The fact that in some instances they not only brought a lump in my throat but made me watery-eyed as well, I do hope I can be forgiven for blowing a few notes on the very ordinary trumpet with which I was equipped at birth.

Such as:—

"A book about Baulking Green is never out of season — least of all when it is written by a man who claims he cannot write — is introduced by John Lawrence in a splendid preface and concerns one man and his horse who between them created a legend in their own lifetime.

The story is told in a good crisp prose like the rattle of hoofs down a lane on a fine September morning when the dew is on the ground and the air is so crisp and sharp that it catches your breath.

Here in this book are the yeomen of England farming their own land, racing and breeding their own horses, caring for their own villages, looking after their own people, tending their own churches and minding their own business. Envying no one, hurting no one.

Ron Liddiard has captured the essence of their quality — and for that alone his is a book to cherish.
John Bliss, "Sporting Life" September 7th 1971

"One would need to have the heart of a stone not to be disarmed by a preface which begins "This is intended to be a country story of a country horse written in simple country language by one who knows no other form of expression". In fact any reader who takes the statement at its face value is in for an exceedingly pleasant surprise. This is a fascinating story written in a fascinating way. — Simple country language be damned! This is the stuff of poetry and Ron Liddiard's book is full of it. Baulking Green may not have been a horse of the stature of Arkle or Persian War, but I found this book infinitely more satisfying than the two books which have been written about that illustrious pair.

Ron Liddiard says that this is the only effort he is ever likely to make to write a book. To this I can only say, I hope not.
M.W. "Light Horse" September 1971

"The author for whom this book is a first-time, writes in a relaxed and natural style, painting a picture of the Old Berkshire countryside and all the people who became involved with Baulking Green, which makes delightful reading. He has captured the spirit of the people who take part in the hunter chasing scene and all those who saw this wonderful horse stretching his neck in a tight finish will now be able to relive his races through these pages.

Wayfarer. "Horse and Hound" October 1971

"Liddiard on a Winner"

In his day Baulking Green was the Arkle of hunter chasing and he could fairly lay claim to being one of the most popular 'character' horses of the last decade.

Ron Liddiard does his subject more than justice in brilliantly simple yet never dull style.

"Sheffield Morning Telegraph" November 4th 1971

While on the subject of Baulking Green the horse, I am reminded of a bit of fun that Jim Reade, myself and Paddy Hogan who came to Faringdon to take part in the army manoeuvre based on sites at Wickwood and Wicklesham in the mid thirties and later returned to marry Kathleen Carter and join her in running the confectionery business next door to Lloyds Bank in the Market Place, when he was demobbed after the second world war.

The three of us decided one evening in the bar of the "Salutation" (the 'Sally' to locals) to enter the fancy dress parade and competition organised by the Faringdon Carnival & Confetti Battle Committee. We hired a pantomime cow costume complete with udder etc and entered her as "Baulking Daisy". A toss of a coin decided that I should be the front half and Paddy the rear-end and that Jim would lead us by a halter wearing his great grandfather's real linen smock which was beautifully smocked on the neck and shoulder yoke depicting that he was the Gaffer — top of the range so to speak. Jim also wore an old milking hat and carried a bucket and stool. After parading round the town we arrived in the Market Place which was packed with on-lookers who enjoyed our aggressive head down charge made at the local police sergeant which dislodged his helmet and sent it rolling down the middle of the square. Unfortunately and perhaps understandably he was not amused and having restored it to its proper place, shouted into the wire mesh through which I was breathing that if we didn't behave ourselves he'd "shop" the three of us.

With that little incident over we lined up with the rest of the entrants in front of Portwell to be judged by Mrs Geoffrey Berners of Little Coxwell. To our surprise we were moved into third place, from which position Jim was asked by Mrs Berners if Daisy was in — milk? "Oh yes Mam" says Jim sitting down immediately on his milking stool with bucket between his legs in the customary manner and shouting "Come on Daisy let 'em see what you can do" whereupon Paddy proceeded to squirt milk into Jim's bucket from a couple of water pistols loaded with milk which Paddy had tied to his belt before disappearing some while earlier into the uncomfortable confines of Daisy's rear-end. In fact Mrs Berners was so impressed with what she had seen that she moved us up a place and awarded us second prize. To round off the episode, accompanied by a brass band we marched in procession to the field in front of Faringdon House from where we were directed to proceed over a narrow plank, bridging the ha-ha on to the lawn. I well remember Jim warning us as we approached this hazard and shouting to us to keep our knees together.

Later in the day half a dozen of us visited the clay pigeon shoot being run for the good of the cause by David Lansdown who had recently taken over Frank Lane's Ironmonger & Agricultural Engineering business in Marlborough Street. To the best of my recollection we had a sweepstake on our combined entry — the winner to take all and the lowest number of 'hits' paid for entrance fees and cartridges. As I am a very poor shot, this surely would have been me but for a few quiet words of persuasion to George Prior the trap operator to throw my old trilby hat into the air in place of a "clay" which enticed Jim Reade to let fly at it with a left and right barrel both of which hit the target but left him short of ammunition to get out of footing the bill, but he did have the satisfaction of drilling a few holes in my old hat and causing a bit of fun.

Having regressed a little I return to the late sixties following Baulking Green's last appearance on a race course at the age of sixteen years.

This period had contributed a great deal of interest and enjoyment in my life up to then but something else was destined to change the pattern we as a family accepted and certainly expected to pursue during our respective journeys towards retirement.

THIRTEEN
Fuller's Earth

It was in the summer of 1968 that I was turning some hay in a field named Middle Home Ground, when a gentleman approached me introducing himself as Dr Brian Kelk from the National Institute of Geological Sciences located in Princes Street, London. He immediately struck me as a very interesting friendly person and explained briefly that the Institute were arranging to carry out a geological survey in the Vale of the White Horse for the purpose of providing a more detailed map of the various soil structures in the area. This exercise would necessitate drilling bore-holes to a depth of around 600 feet in order to provide sufficient soil cores to take away for laboratory analysis.

Dr Kelk also informed me of other sites, being investigated and he would appreciate permission to put down a bore-hole in a position he indicated on part of an ordinance survey map, adding that it would probably be done in about six or seven months time, say next February. I would be granted an emolument of £5 and the site would be left clean and tidy. After explaining that the chosen spot would be difficult to reach on our wet old land at that time of year Dr Kelk said it mattered little within a few hundred yards so "how about in that gateway adjoining the public road and if we need to extend the area with a little broken stone to accommodate the lorry housing the rig and the caravan accommodation for the two operators, that would be fine". I mention this because in view of the discovery later of a sizeable deposit of calcium montmorillonite, known as Fuller's Earth — this would not have been found if the drilling had taken place on the spot originally suggested for the reason that it would have been on the perimeter of the deposit and would have failed to indicate its size and importance, as well as missing Bed 1 altogether, which contained samples with a purity of over 90%. In fact "Borehole 1" as it became known turned out to be almost bang in the centre of the deposit.

This initial drilling was followed some months later by blanket drilling the area which indicated a sizeable amount of Fuller's Earth lay under approximately 120 acres of land situated mainly on farms belonging to Jim Matthews (Spencers Farm) Jim Reade (Manor Farm) and myself (Church Farm).

The eventual publication of the comprehensive report of the exercise and its results by the I.G.S. lead to a great deal of interest being shown by a number of firms, seventeen in all, who put forward various offers and suggestions as to how the deposit could be excavated. A great deal of local publicity was given to the discovery and there was a lot of exaggerated talk about possible fortunes to be made.

Understandably the 50 or so people living in the village some of them representing three or four generations were apprehensive about the effect of nearby quarrying and the drying and processing of the materials in a small factory type plant situated about half a mile from the cluster of dwellings which mostly edge the western boundary of the village green.

There were others in neighbouring villages who were naturally concerned about increased lorry traffic becoming a problem.

The other side of the coin revealed that the discovery of the deposit being in excess of 500,000 tons was of national importance as the two or three sites being worked were running out of materials and there were very few known sites suitable for exploration. The material has many uses, since it was originally discovered and used many years ago for "fulling" cloth. Its main use now being in the foundry industry where it is used extensively for the bonding of sand in moulds for casting metals. It is exported in large quantities to Finland and Sweden for use in their paper making industry. It also has uses in oil-well drilling both as a lubricant and in refining the finished product, as well as specific uses in civil engineering in connection with pile driving and bridge building.

What I have said perhaps conveys the more general situation in which we found ourselves. It was a situation that provoked a great deal of heart searching and while I am sure some of the thoughts that passed through the minds of Jim Matthews and Jim Reade were similar in some respects to mine, our circumstances were very different. I was also very conscious indeed of how important a decision to become part of the development would be to the future of the rest of the family. It was on this basis that, after a great deal of deliberation, a decision to participate in the development and excavation of the deposit was reached. Having made up our minds I think it is fair to say we all knew we were destined to sink or swim together.

The effect meant that in the first year or two of the estimated twenty year life of the quarrying, with roughly a further 85 acres unaffected by the working it might be possible to retain the dairy herd and devise a new grazing plan for them, which would of course need altering frequently as quarrying and restoration progressed.

While this may have been possible to start with, the real difficulties were bound to arise when the farm and dairy buildings became almost isolated by quarry working and the unknown factors connected with land restoration such as when it was likely to be ready to carry stock once again. It was realised this would be far to risky so it was sadly decided to give up milking; sell the cows; take advantage of the Ministry of Agriculture's grant scheme to encourage farmers to give up milking, in an effort to solve the problem of over-production, and convert the cow-housing area into a calf rearing unit buying in batches of beef calves for sale as stores weighing around 6/7 cwt a piece. This was accompanied by a small bull-beef unit which while being quite an interesting project was, I quite often thought, fraught with a certain amount of danger when the young bulls approached the age of puberty. Fortunately, I had more or less retired by then so Bill with his willing worker "Inch" Lawrence were the ones who faced danger during 'littering up' with straw when necessary two or three times a week, when I'm sure another pair of eyes in their backsides would have been an advantage.

To complete the farming plan the land surplus to stock and conservation requirements would be planted to wheat and winter barley, together with any inaccessible areas caused by quarrying.

By now we had formed a small company with Jim Reade and Jim Matthews putting our respective Fuller's Earth bearing areas into an entity for the purpose of negotiating with those interested in developing the deposit.

We were informed it would take roughly ten years from the time of discovery to reach production which proved a very accurate forecast. The fact that excavation started on the northern perimeter of the deposit on Jim Reade's land in the summer of 1978 meant that we had a further two or three years before quarrying reached our land giving us more time to sort ourselves out long term. It had become increasingly clear to us that eventually Church Farm would no longer be a viable farming unit and incapable of providing a living for Bill, his wife Trish and their three children Sally, Kate and James. So it was that a search began through agencies and in the farming press for a suitable medium sized dairy holding preferably somewhere in central, south or western counties. In fact Bill and Trish visited in excess of forty such properties in Cornwall, Devon, Somerset, Wiltshire, Hereford, Shropshire and Northampton which provided only about five "possible" out of which two reached the serious stage involving offers which could only be made by juggling a few assets and praying for a following wind — neither of which amounted to enough in either case.

I'm sure this turned out to be somewhat of a blessing in disguise, as in December 1987 applications were invited for the tenancy of a 230 acre mixed dairy and arable farm in North Hampshire being part of the Manydown Estate belonging to the Oliver-Bellasis family who very fortunately for us selected Bill from a sizeable number of applicants and he commenced farming on his own account at Skyers Farm, Ramsdell on December 28th 1987.

I would like to think that after ten years his landlord is reasonably happy with his tenant, and although I say, as perhaps shouldn't, that this may be so if any measure of achievement can be judged by the fact that ably assisted by Trish he managed to win the award for the best dairy herd in the Basingstoke Agricultural Society's Show in 1996 and was placed third in the best farmed farm under 400 acres, both classes having ten or more entrants.

Becoming a tenant instead of an owner-occupier as anticipated earlier meant unscrambling our original plans, devising a fresh approach, retaining a few assets but parting with the old homestead of Church Farm itself with about five acres of adjoining land. I had lived there since I was six years old, with mother and father, taken Ann there as my wife in 1939, saw Pam married from there and Bill take it over as his home with Trish when their third child James was born in 1977. We changed houses with them to live where they started their married life at Forty's, a cottage which had served up to the turn of the 18th Century as a farmhouse to a small holding described on the 1912 Ordinance Survey as Forty's Farm.

Living at Church Farm for almost sixty years I naturally developed a love of the old place and often wondered about the people who had lived and worked there before me, some of whom had left their initials branded on the wooden beam set above the old open fireplace in the back kitchen. The initials G.C. & R.W. representing George Collins and Robert Whitfield were very clear and no doubt branded by a red hot branding iron heated in the open fire below for the purpose of indicating who was the rightful owner of small hand tools such as two and four grained prongs, spades, axes, billhooks and the like.

There were a small number of friends and professionals whose help and support while setting up the Fuller's Earth project was greatly appreciated. I don't propose to name names but if per chance any of them happen to read this they will know and also that I am eternally grateful to each of them.

Now that the extraction of the deposit is nearing its end it is possible to look back on the 18 years that have passed since operations commenced. Taking into account the difficulty of removing an enormous amount of top soil and overburden to take out some 500,00 tonnes of Fuller's Earth from depths up to a maximum of 90 feet, close to a small village community, processing it on site and transporting the end product at home and abroad, was no mean task to say the least. Nevertheless, I think it is fair to say the operators have shown a high level of consideration and done all they could to minimise as much interference as possible, as well as providing a couple of dozen jobs in the locality and a considerable amount of business such as transport, brought about by a knock-on effect.

FOURTEEN
Tailings

I count myself extremely lucky that by the Grace of God and a following wind I have reached my 85th year. Considering the accepted "par" for the course is three score and ten I reckon I'm on a generous bonus and often relax in the armchair to count my blessings.

During the journey I have been more than fortunate in meeting countless lovely people from varied walks of life, and done my best to adopt a reasonable moral philosophy. I have also come to regard a few locations, poems and sayings as being a bit special. I think most of us have our own ideas and particular favourites; for instance I never tire of reading Rudyard Kiplings "If" and have derived a great deal of help and comfort from it in times of trouble and uncertainty.

Only very recently I had the good fortune to be given a book of poems to read by Wilfred Howe-Nurse entitled "Berkshire Vale". They are beautifully illustrated by Cecil Aldin and were published by Blackwells in 1926. The book was lent to me by Sir John Betjeman's daughter Candida Lycett-Green who said they were among some of her father's favourite poems. No mean recommendation, coming from a past Poet Laureate, they are so descriptive of the Vale of the White Horse which he came to love so much living at Uffington in the 1930's. The two short poems which concerned White Horse Hill and Rosey Brook are the poems which appeal to me most, both being favourite spots which over the years seem in a strange way to portray a sense of wonderment and perspective which have contributed towards the make-up of my life in the Vale.

Occasionally Ann and I would sit in the car on top of the hill to relax and enjoy the view northwards across the Thames Valley to the Cotswolds: and I don't find it all that difficult to picture the 'goings on' during the Annual Scourings of the White Horse when it was recorded that in 1780 the event attracted 30,000 people who took part or witnessed all kinds of foot and cart horse races supported by competitions including wrestling and back-swording, the contestants for which coming from most of the southern counties and as far west as Somerset and Devon. It is said that the Somerset men contesting in backswording drank as much vinegar as they could prior to taking part, as it helped to delay bleeding, giving them more chance of success, as the outcome of the contest was decided basically on the amount of blood from wounds inflicted by the ash sticks used as swords. With grateful acknowledgement to Wilfred Howe-Nurse the words of his short poem entitled V.W.H. are as follows:—

> Stand on White Horse Hill,
> Watch the Vale below
> When all the world is still,
> So thou mayst know
>
> How brief is mortals span,
> How weak the ruler's rod;
> How small the works of man
> How great the works of God.

I wonder if those two short verses warrant recording on a suitable tablet of some kind situated perhaps close to the car park for the disabled, near the top of the hill?

The second of my favourite spots, Rosey Brook became so for the fascination and fun it provided for me in so many ways. It is a long time since I strolled along its banks from Moor Mill past the double home-made withy gates guarding each side of the Long Lane bridge constructed of old railway sleepers, strong enough to take a team of cart horses and a wagon load of hay, or withy poles cut in the Withy Bed on what was known as "the Fernham side of the brook". Grampy Ernest once owned the Withy Bed and when father first came to Church Farm he was given permission to cut and fetch one load of poles a year free of charge and there was usually a sigh of relief from the carter when the load had crossed the bridge on to "the Baulking side". The brook-side stroll proceeded past the "bathing pool" and the spot where Mr Loyd-Thomas is reported to have jumped the brook while out hunting in the 1930's with a leap that appeared to border on the impossible when I last stood and gazed at the site. It then snakes its way through Chestnut Walk and the "Ford" in which I had great difficulty in preventing an old pony lying down with me when most of the hunting field were splashing through it when the brook was in flood and impossible to define, let alone jump. My stroll usually left the brook close to Rosey Covert at the point where the foot-bridge accommodates the old foot-path between Baulking and Shellingford, and if one follows this path southwards it joins Baulking village green close to Hyde Farm, often referred to as Lower Farm in the same way as Church Farm was defined on the Ordinance map as Upper Farm, the probable reason being that both farms at one time belonged to the Hyde Estate which I believe was centred at Hyde End near to Brimpton and Newbury.

My first memories of Rosey Brook were formed when I was about ten years old. It was a favourite area in which to play "Fox and Hounds" which was akin to taking part in a paper chase, without dropping paper. Usually word got around the teenagers in the village during the lighter evenings that a gathering for the purpose was taking place and from the two dozen or so who turned up, two were selected to be foxes and given about 15 minutes start, while the rest being the hounds then set off accompanied by imitations of hounds in full cry to find their quarries and then by sight hunt them down.

When I reached my teens the brook provided the fascination

of crayfishing by the light of the harvest moon. In those days there was an abundance of crayfish in the brook and a haul of around 150 mature fish was common place with a large number of younger ones being thrown back. Alas during the 1950's there was a sharp decline in numbers, due it was suggested to the increased use of tarmac on our roads resulting in pollution of brooks by surface drainage. I have been told that a recent survey has revealed there are no crayfish at all in this stretch of Rosey Brook and the last record of any being caught was in 1972. Coupled with lasting memories of seeing the Old Berks hounds hunting this area from a vantage point on Baulking Hill and the connections it has with my introduction to the sport of horse racing "Between the Flags" makes it special and prompts me to quote the three touching verses by Wilfred Howe-Nurse entitled:—

Rosey Brook

The whispering reeds of Rosey Brook
in summer sing a happy tune
Of love and joy and happiness
and all the lazy life of June.

But when grey winter calls the song
they moan, like stricken souls in pain.
A requiem for bygone days
that never can return again.

Yet if you walk by Rosey Brook
at evening, from the world apart,
They sing of peace, and you may hear
their song re-echo in your heart.

So much for my two favourite spots. The following are from a few sayings jotted down over the years upon hearing or reading:—

"Sleep is for the body
Silence for the soul."

"Courage grows with daring
Fear with holding back."

"To-day is the tomorrow,
you worried about yesterday
and all is well."

"A man who NEVER makes mistakes
NEVER makes anything."

"Work faithfully 8 hours a day and don't worry
Then in time you may become the boss
Work 12 hours a day and have all the worry."

The Farmer's Lot

We are ruined if it rains
We are ruined if it doesn't
If the crops are good the prices are bad
And if the crops are poor we've nothing to sell.

The following was told to me by Captain Tim Forster who among hundreds of other winners trained "Well to do", "Ben Nevis" and "Last Suspect" to win three Grand Nationals.

He said "Bing" Lowe who farmed Fox Farm near Stow-on-the-Wold, and hosted the Heythrop Point to Point fixture for many years told him "There were only two necessary things in life — one was a good bed and the other was a good pair of boots — as you were likely to spend half of your life in both."

I came across the following Ode to Laker which I cut from the "Daily Mirror" in 1956. It was composed by two brothers aged 12 and 10 years old to commemorate the taking by Jim

Laker of all ten Australian wickets in one Test Match innings. It earned the composers one guinea each from the Sports Editor and was entitled:—

"Howzat"

TEN little Aussie boys Lakered in a line
Harvey caught by Cowdrey and then there were nine
NINE little Aussie boys now full of hate
Burke was the next to go and then there were eight
EIGHT little Aussie boys praying up to Heaven
Craig was l.b.w. and then there were seven
SEVEN little Aussie boys all in a fix
Mackay was caught by Oakman and then there were six
SIX little Aussie boys batting was a strive
Miller bowled by Laker and then there were five
FIVE little Aussie boys runs were looking poor
Archer caught by Oakman and then there were four
FOUR little Aussie boys come in after tea
McDonald caught by Oakman and then there were three
THREE little Aussie boys feeling rather blue
Benaud bowled by Laker and then there were two
TWO little Aussie boys looking very glum
Lindwall caught by Lock and then there was one
ONE little Aussie boy not having fun
Maddocks l.b.w. and then there was NONE.

At the other end of the scale, I think most of us have times when perhaps we "bow the head and bend the knee a little," for the purpose of doing a bit of soul searching. It is then that the following words given to me by Mrs. Seton Lloyd of Woolstone, following a discussion group meeting in her house, form the basis of my thoughts and seem to point towards the way in which everyone of us could help to build a better world to live in:—

"We are all agents of God. Every day we touch countless lives in one way or another. How close we are to God in consciousness at any given moment will determine whether or not we are being loving and constructive in our relations with others".

The words mean as much to me now as they did then, except that their regular use reminds me what a poor Christian I really am and furthermore what earnest praying is all about.

From religion to realism. Apart from the expected I often think it is amazing what can happen in a day. For instance Ann and I woke up to the frosty morning of February 3rd 1996 to receive a 'phone call at around 11 am from Candida Lycett-Green inquiring if our little Church of St Nicholas was locked. She explained that as the ground was too frost-bound for hunting, she was showing H.R.H. The Prince of Wales around some of the villages and churches in the White Horse Vale and could she bring him to our modest abode for a short chat about our village in about half an hour's time. What an honour it was to welcome him with a little drop of "Gin & Ginger" which he seemed to enjoy in an extremely relaxed manner free of formality and fuss before leaving his signature in our visitors book as well as at the Church, as an everlasting reminder of one of the highlights of our lives.

It is fair to say that what happened to us on May 26th 1993 falls into a very different category. While it was also very unexpected, it would have been a great deal better if, it had not happened at all. Going down to make a cup of tea that morning I was faced at the bottom of the stairs with the sight of the hall carpet floating on 9 inches of water, due to torrential rain falling during a violent thunderstorm in the early hours causing flash flooding and considerable damage throughout the Vale.

Evidence of the damage can be gauged from the accompanying photographs and the fact that it was a fortnight before the dehydration units were able to dry the house out and new carpets and curtains etc., were fitted to enable us with the help and consideration of the N.F.U. Mutual Insurance Society and a very caring family circle we were able to take up residence once more.

Baulking has seen a great deal of change since I came to live here 78 years ago. Due to great steps forward in mechanisation, farming is no longer a majority employer in the village. The same area of land is now farmed by about one tenth of the former work force. Farmhouses and cottages have been modernised and in a few cases barns and cattle sheds have been converted to desirable dwellings to house a very different but still small village population who are engaged in widely varying occupations.

It is inevitable that such changes leave a trail of memories in the minds of those who experienced them. Quite recently I was given a poem which expresses a sentiment I feel is relative to the period.

The Good Old Days

We met, we married, a long time ago
We worked for long hours and wages were low.
No telly, no radio, no bath, times were hard,
just a cold water tap and a walk up the yard.
No holidays abroad, no carpets on floors,
We had coal on the fire, we never locked doors.
Our children arrived, no pill in those days,
And we brought them up without the State aid.
No valium, no drugs, no LSD,
We cured our pains with a good cup of tea.
If you were sick, you were treated at once,
Not "fill in a form and come back next month".
No vandals, no muggings, there was nowt to rob,
In fact, you were rich with a couple of bob.
People were happier in those far off days,
Kinder and caring in so many ways.
Milkmen and paper boys would whistle and sing,
And a night at the flicks a wonderful thing.
Oh, we had our share of troubles and strife,
But we had to face it, that was life.
But now I'm alone and look back through the years,
I don't think of bad times, the troubles and tears.

I remember the blessings, our home and our love,
We shared them together and for those I thank God above!

    I have tried to find out who wrote this poem but, alas, without success. It is reproduced here with a very sincere acknowledgement to the anonymous author.

    While writing the last few words of something I started for a bit of fun ten years ago, the sight of a fossil adorning the hearth around the fireplace in our sitting room has the immediate effect of putting it all into perspective and reducing my effort to its proper proportion.

    The fossil itself weighs about 10½lbs and is the size and shape of a rugger ball. It had been laying buried some fourteen feet below ground level and was discovered when a seam of Fuller's Earth was being excavated for the purpose of a bulk field trial prior to developing the deposit.

    Anxious to learn something about it I took it to the Geological Department of Oxford University where I was told it was a marine-type snail named a Nautilus.

*'What's New Pussycat!'—A Fossilised Nautilus
Believed to be one hundred million years old.*

The answer to my enquiry as to how it got there, was that millions of years ago there was a fast flowing river in the area and when it eventually subsided some of these creatures became trapped in pools left behind, died and became fossilised by the natural chemical reaction activated by the various ingredients surrounding them.

However the biggest surprise of all came when I asked my interesting and helpful informant if he could hazard a guess as to how old this particular Nautilus was. He replied that the lower crustaceous age started about 125 million years ago and ended some 50 million years later, so give or take a few million either way it was probably in the region of 100 million years old.

Sometimes as I sit in my armchair gazing at this strange object which swam around Baulking so long ago, my mind is filled with all sorts of fascinating thoughts which are far beyond my powers of understanding. Nevertheless I am always left with a profound feeling that there MUST be a purpose behind it all and with apologies to William Shakespeare "we all play our parts as we flit momentarily across the stage and are gone".

This leaves me with nothing but the sincere hope that I have sufficient faith in me to be recognised in the mannner illustrated by a few lines I saw pinned to the back of a church door in North Wales which said quite simply:—

> Each time I pass this Church
> I pay a little visit
> So when at last I'm carried in
> The Lord won't say, "Who is it?"

FINIS